The Basics of LiveStreaming

By Paul Richards

Copyright © 2021 Paul Richards

All rights reserved.

ISBN: 9798557634410

DEDICATION

To the communicators. To the innovators. And to those who are working to connect the dots and make it happen.

CONTENTS

	Acknowledgments	i
1	Introduction	1
2	Brief history of livestreaming	Pg 4
3	What do I need to livestream?	Pg 8
5	Ten tips for hosting a great livestream	Pg 14
5	How to make an engaging livestream	Pg 17
6	How to livestream to YouTube	Pg 22
7	How to livestream to Facebook	Pg 27
8	How to livestream a Zoom meeting	Pg 31
9	What is OBS	Pg 35
10	What is vMix?	Pg 41
11	What is the best camera for livestreaming?	Pg 46
12	How much bandwidth do I need?	Pg 56
13	What type of computer do I need?	Pg 63
14	What type of cables do I need?	Pg 71
15	How to add graphics to your live stream	Pg 82
16	What is NDI	Pg 91
17	What is a PTZ camera?	Pg 94
18	What is a SDI camera?	Pg 97
19	What is an NDI camera?	Pg 100
20	What is a Tally Light?	Pg 105

THE BASICS OF LIVESTREAMING

21 How to build a livestreaming studio?   Pg 109

22 What is SRT?   Pg 114

23 Conclusion   Pg 119

ACKNOWLEDGMENTS

Tom Sinclair from the Streaming Idiots for your endless streams of educational information. Tim Vandenburg from vMix, for your educational content. Finally, Michael Luttermoser and Tess Protesto on the StreamGeeks team.

1. INTRODUCTION

The *Basics of Livestreaming* is an introductory online course and book that will teach you the essentials of live video streaming. The course and book offer learning opportunities for anyone who is interested in audio/visual equipment and streaming technology.

This material offers a good foundation for more advanced online courses from the StreamGeeks and outlines skills that can be applied to any streaming media project. For example, *The Basics of LiveStreaming* course is the ideal primer before taking the OBS, vMix, or Wirecast courses that are available on Udemy. The course is also ideal for volunteers who operate livestreaming equipment and aspiring video producers.

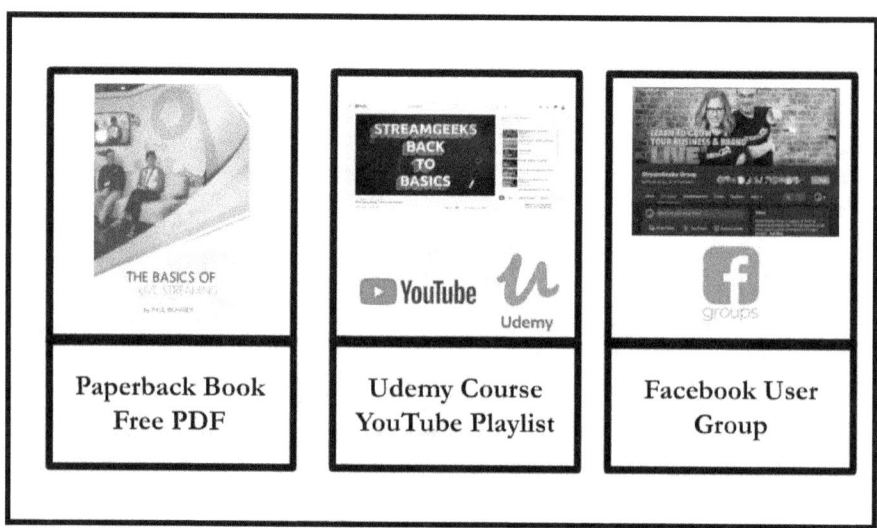

Additional learning opportunities.

This book will help you get the most out of livestreaming and open your eyes to exciting new opportunities and use cases in streaming media. Each chapter is accompanied by a video tutorial. Our production head, Michael Luttermoser, detailed each chapter with video to accompany the book.

Author Paul Richards and StreamGeeks Producer Michael Luttermoser.

Each video has a YouTube playlist which provides a simple format for those who want to "skip ahead" to watch only the subjects that interest them. Inside each video description, you will find a video index which will enable you to find the specific sections of information you want on YouTube.

THE BASICS OF LIVESTREAMING

Outside the 2019 StreamGeeks Summit, Dream Downtown Hotel NYC.

Finally, some segments of this course are enhanced by the free Udemy course. Here, you will find additional digital materials, quizzes, and reference links. If you take the course on Udemy you will also receive a certification that can be applied to your resume and LinkedIn profile.

The StreamGeeks are happy to answer your questions from inside our Facebook User Group. Post questions and get feedback from the entire StreamGeeks community by engaging with the group.

Online Group Link:

https://www.facebook.com/groups/streamgeeks

2. A BRIEF HISTORY OF LIVESTREAMING

Live streaming as we know it today has been enabled by a few important technological advancements -- let's start with computers. Each year, computers become faster and more affordable, opening up new markets of technology users every day. The second advancement is the internet which has enabled and expanded the use of livestreaming. Each year, access to the internet increases, allowing streaming video to reach billions of people with better quality and reliability.

Analog and digital video compression standards were being developed long before the internet. In the early days of livestreaming, the primary issue with livestreaming video was encoding algorithms. Engineers needed to find a way to compress video so that they could send it between computers using cabling with limited bandwidth capabilities. Early video codec engineers looked for ways of transporting video that was both efficient (low bandwidth) and high quality.

In order to send and receive "streaming" video, one needs to first compress or encode the video, but also decode the video on the video receiving with a computer. Early digital video compression algorithms required bitrates that were unrealistic for distribution over public communications infrastructure. A "bitrate" is a measure used to determine the amount of data per second that is being sent for livestreaming audio and video.

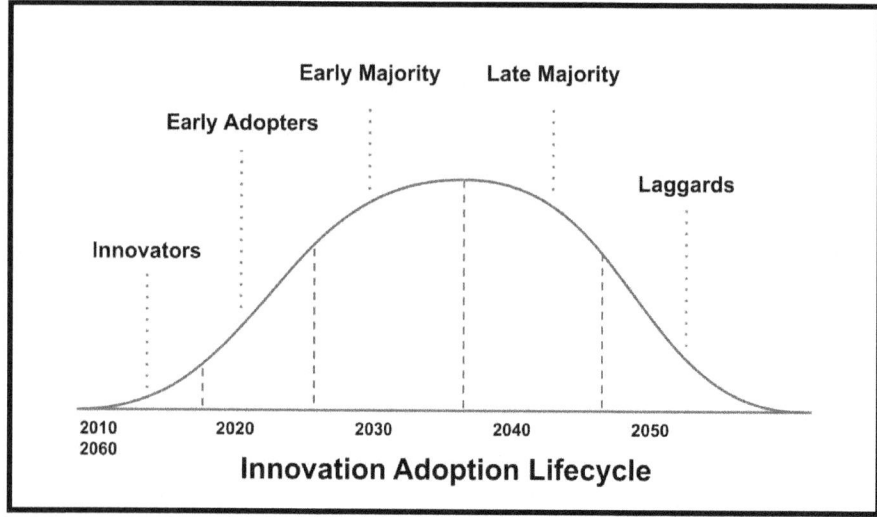

Technology adoption bell curve diagram.

As video encoding technologies became more efficient, computers also became faster. The earliest forms of live streaming video happened on local area networks (LANs) between computers connected in a single room or building. As internet access expanded, consumers started to gain enough access to bandwidth to watch live video online. In 1995, the first baseball game was live streamed and by 1999, Bill Clinton became the first U.S. president to participate in a webcast. In 2005, YouTube launched and the world of online video really began to take shape. Five years later, in 2010, YouTube was used to livestream a message from President Barack Obama. By 2013, YouTube officially opened up live streaming to the public by releasing YouTube Live Studio.

The first iPhone was released in 2007, but it wasn't until 2014 that companies started building applications that could deliver live video to smartphones. Periscope was a company that developed a live streaming app for iOS and Android phones in 2014. Periscope was acquired by Twitter in 2015, and the application brought live streaming video to the masses with an easy-to-use smartphone-friendly application. In 2016, Facebook announced that it would bring live streaming to its platform. This move brought livestreaming to over 1 billion Facebook users in a matter of months.

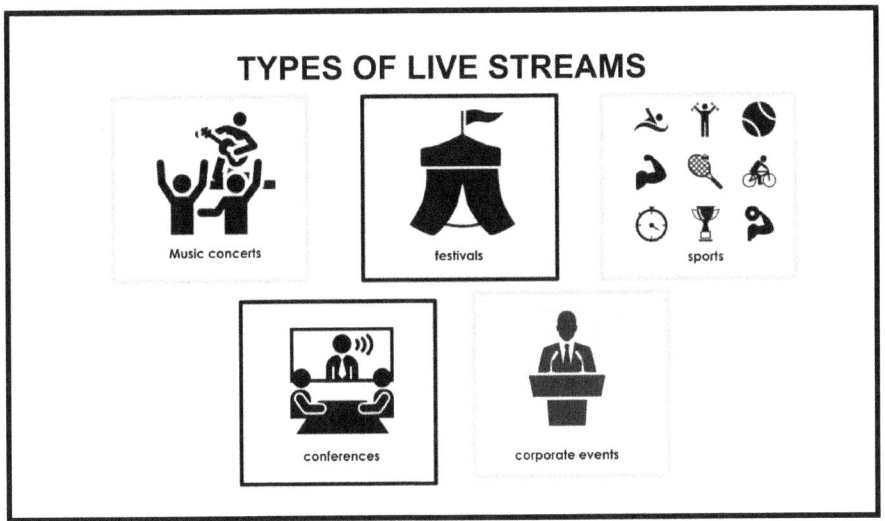

A few types of livestreaming events.

From meager beginnings, online video streaming started growing at an astonishing rate around the world. As access to computers, smartphones, and internet bandwidth spread, the need for consumer access to livestreaming content grew year over year. Simultaneously, video communications and on-demand video grew as well. Video conferencing companies like Zoom, Webex, and GoToMeeting have emerged as leaders in online communications, while Netflix, Amazon, and Disney thrive as providers of premium video content. . Companies like Zoom allow users to livestream their meetings to social media websites such as Facebook and YouTube which you will learn about in an upcoming chapter.

The market of online video is set to grow at outstanding rates in the time between 2020 and 2030. IThis book will help you better understand the basics of livestreaming technology and where it's headed in this decade. I believe there's never been a better time to learn about livestreaming technology and implement a strategy for your business, organization, or personal projects.

3. WHAT DO I NEED TO LIVESTREAM?

Livestreaming is a process of delivering audio and video to audiences using the internet. It is very easy to livestream using a smartphone with an app such as Facebook or YouTube. In this chapter, you will learn the fundamental concepts of online video streaming.

In order to livestream, you will need an internet connection, audio and video sources, an encoder, and access to a streaming destination.

What is a content delivery network?

Streaming destinations, also known as content delivery networks, receive a stream from your encoder and distribute it to large audiences. Popular free content delivery networks (CDNs) include Facebook, YouTube, Twitch, and LinkedIn. Private content delivery networks allow you to charge for access to your livestream. They also provide other premium services such as branding and video replay. Popular private content delivery networks include Vimeo, DaCast, and StreamShark.

THE BASICS OF LIVESTREAMING

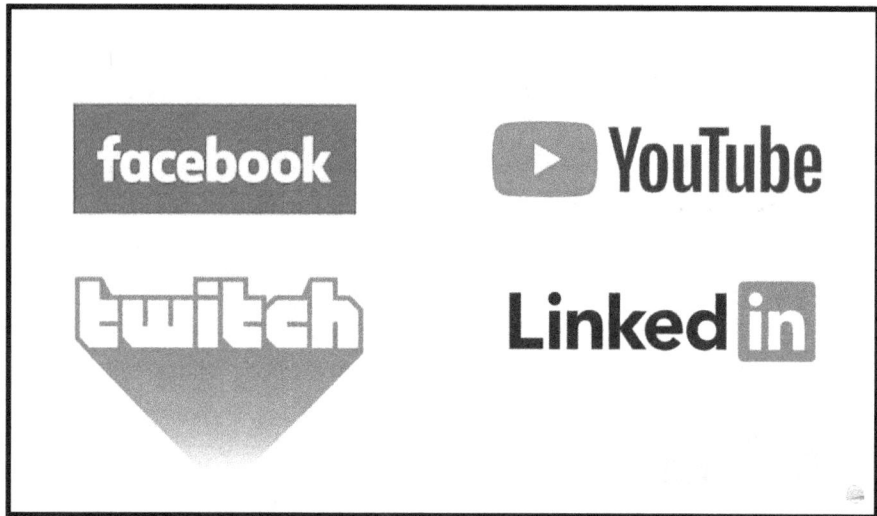

Examples of Content Delivery Networks that allow you to livestream to large audiences for free.

All CDNs provide users with a streaming URL and a secret key. You can retrieve this information from the CDN's website and enter it into your encoder software or hardware system. This is how your encoder is able to livestream directly to the CDN of your choice using the internet.

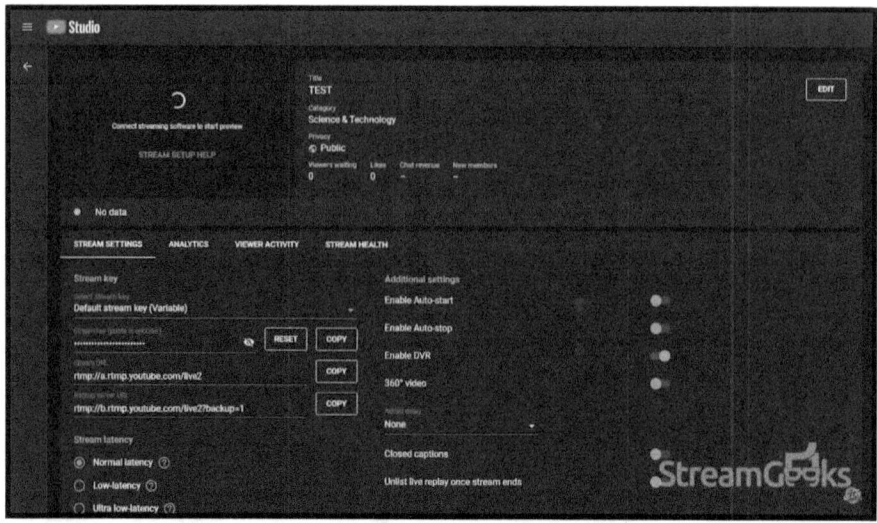

Example of a livestream test on YouTube's livestreaming dashboard.

What is an encoder?

Your smartphone can become an encoder with the help of a livestreaming-enabled app. An encoder takes audio and video sources and encodes them into a stream for CDNs. Encoders use compression to combine audio and video into a reliable stream. There are software and hardware encoders available for livestreaming. Software encoders include Open Broadcaster Software (OBS), Wirecast, or vMix. Hardware encoders are physical devices that are plugged into cameras and audio mixers in order to produce a live- stream.

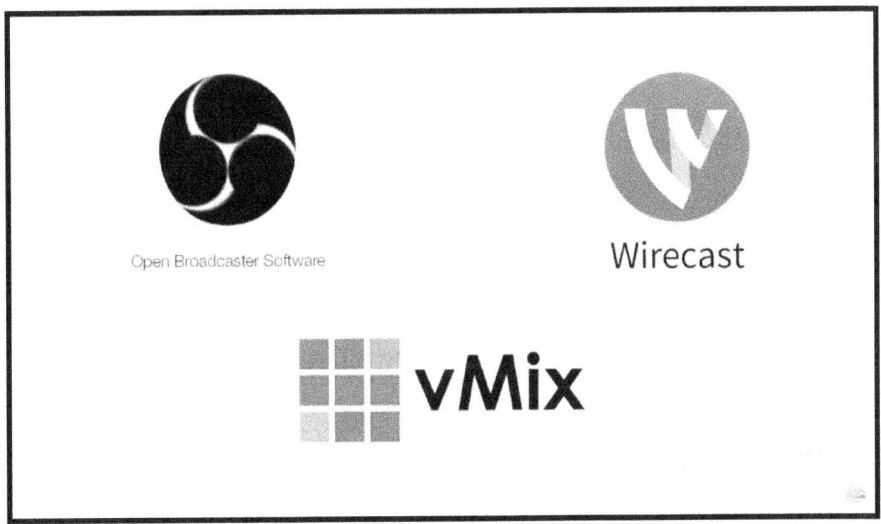

Popular livestreaming software.

What is Real-Time Messaging Protocol and what bitrate should I use?

When you live stream, you're using your upload bandwidth to send data to a CDN. Most live- streams use a data transport method called Real-Time Messaging Protocol (RTMP) which can be used to compress your audio and video sources using various bitrates. The higher you set your bitrate, the more data you're able to send. Also, the higher your bitrate, the higher the quality of your livestream.

Generally, you should never use more than half of your total upload speeds for your RTMP live stream. A 4-6 Mbps bitrate (stream quality) is considered television quality. This is about the bitrate you will see on most Netflix or Amazon Prime video content with good internet access. Your bitrate should be set depending on the resolution and framerate of your production. The higher the resolution and frame rate, the higher your bitrate should be set.

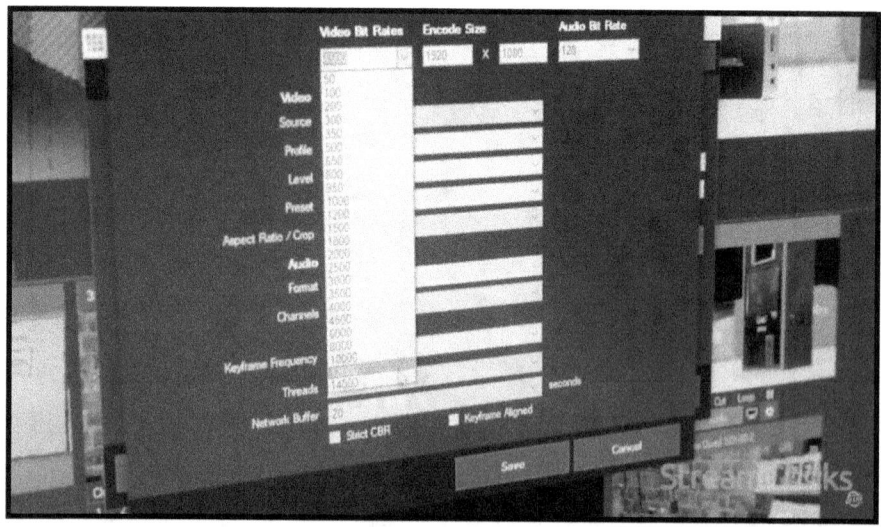

Example of bitrate options for encoding a livestream with vMix.

What are the basic steps to livestream?

The first step is setting up your encoder to work with your audio and video sources. For example, you may want to run

OBS on your laptop. You could decide to use a USB webcam to connect a video source to your computer. You may also decide to use a USB audio interface to connect multiple microphones to OBS. Once you have your audio and video sources connected to OBS, you can retrieve your RTMP streaming information. You can then decide to login to Facebook, YouTube, or Twitch directly through the streaming settings of OBS. Or you can retrieve the RTMP URL and Secret Key from the CDN of your choice and enter that into OBS. Once your streaming destination is configured you can click "Start Streaming" to begin your livestream.

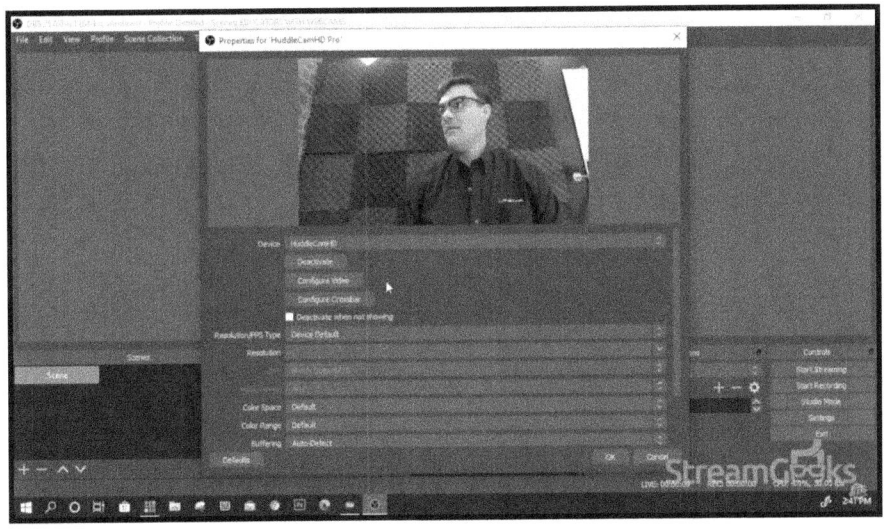

Example use of OBS, the world's most popular free livestreaming software.

Some CDNs like YouTube require you to press a "Start Stream" button inside of their website to officially start a stream. Other services like Twitch and Facebook can be set up to start streaming as soon as they start receiving your livestream.

4. TEN TIPS FOR HOSTING A SUCCESSFUL LIVESTREAM

Okay, it's time to review 10 quick tips for hosting a successful livestream. If you have been following along in our Back to Basics Livestreaming Course, you already know what you need to livestream. You know about cameras, software, social media networks, and audience engagement ideas. Now we'd like to cover some of the most important tips the StreamGeeks have learned about live- streaming over the past couple of years.

#1: Content is King. When you think about hosting a livestream, consider the topic you will cover and the ways that you can deliver valuable information. Keep it fun and interesting for viewers. Spend time creating an outline and pair that outline with visual assets that will bring your ideas to life.

#2: Brand your show. The great thing about a livestream is that you can create fun and engaging content on a regular basis without too much hassle once it's all set up. Take some time in the beginning to brand your show. This includes lower thirds (a graphic used to display information in the lower third of the screen), logo placement, social media titles, and introductory videos. Getting your branding right from the start,will make all of your content look better and your catalog of videos will perform better.

#3: If you are going to be on-camera, get comfortable with it. If you are the producer, then help your talent feel comfortable. At the StreamGeeks, we like to play pump-up music before our livestreams. I even like to jump up and down and stretch a little. Get psyched up, because your audience will feel your energy.

#4: Create a space for your live-streams. At the StreamGeeks we have built multiple studios and they create the inspiration we pull from each show. Yes, you want to have good lighting in this space, cameras, and microphones. But add a style that aligns with your brand.

#5: Get creative with live-streaming and what you put on-camera. Zoom in on details. Show your audience behind-the-scenes views. Push the limits and know it's okay to make mistakes or show a part of you that isn't 100% business.

#6: Schedule your livestreams and show your audience a countdown timer. In many cases, Facebook and YouTube will notify your followers of your livestream via the countdown. Scheduling allows you to get the link to your livestream in advance and share it with followers and guests. Always schedule important livestreams.

#7: As you start to build your audience, network with other professionals in your industry. Check out what other people are doing, and see if they might be good guests for your show. Connect with your viewers on social media, and follow them back. If they are interested in your live show, maybe they can provide you with valuable feedback.

#8: Consider taking the best parts of your livestream and creating shorter, more consumable video segments for social media. Sometimes after a livestream, the StreamGeeks record a livestream recap. This is a great way to deliver the best gems from your livestream and repurpose them for shorter segments.

#9: Look to your audience and followers to supply ideas for new content. Once you gather a large enough audience, listen to the comments and craft new video content and livestream ideas based on audience comments and suggestions

#10: Remember to inject some fun into your livestreams whenever possible. Laughter is contagious, and entertainment thrives on positive energy.

5. HOW TO MAKE AN ENGAGING LIVESTREAM

Engaging video content prompts an audience to participate in the conversation. It's content that's worth sharing and it stimulates curiosity. Engaging content is generally presented in a way that follows basic storytelling rules. Good storytelling allows viewers to imagine that they are the hero of the story. Broadcasters create stories with a compelling hook, an interesting middle, and cliffhanger ending that will keep viewers coming back for more and sharing the content with their social media networks.

Components of storytelling.

How do you find an audience?

For some broadcasters who are just starting out, finding an audience can be challenging but it's certainly not impossible.

The easiest way to create an audience is to think about building a community. Audiences for live video content will build slowly, but for some, they grow quickly because the content goes viral. For most broadcasters, audience-building is a slow process. Consider creating a blog that can help you build your email list. Start a Facebook Group that cultivates shared conversations. Social media offers great tools for building online interest groups that align with the interests of your brand. Use the tools that are available to you, be true to your brand's message, and create content consistently. Don't be afraid to livestream just because you don't have an audience.

The StreamGeeks' Facebook Group.

What is viewer engagement?

Viewers can engage with livestreams in many ways, but the most prominent way is through the chatroom. Engaging video content can stimulate an active chatroom where viewers can interact and contribute to an evolving conversation that is happening in real-time. Engaging content prompts viewers to share comments of support, or question the material that you are presenting. Some livestreams even benefit from viewer engagement that happens after the broadcast in the form of social media posts. Be sure to check your social media feeds and respond to social media posts the same way you would with live chat messages.

A producer engaging with her audience.

How do you promote your livestreams?

Most large social media networks such as YouTube and Facebook do not allow you to promote your live streams on their platforms until after the broadcast. Consider promoting your best performing content as a way to grow your audience online with likes and subscribers. Some platforms such as Amazon do allow you to advertise livestreams that feature clickable links to products that you can promote on Amazon.

Most advertising tools on social media allow you to promote existing content before your livestreams happen. Use these tools to grow your audience and schedule your livestreams as an organic way to notify your audience of upcoming events.

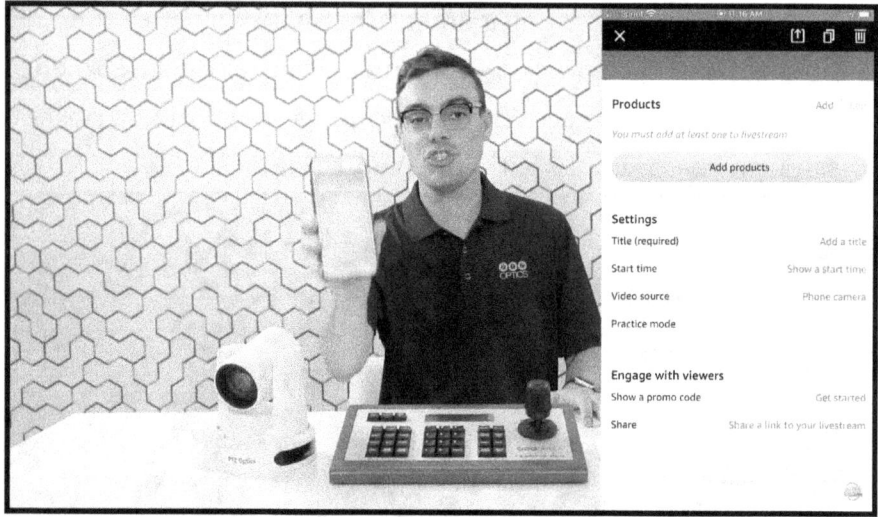

Livestreaming on Amazon Live.

How can you measure successful engagement?

Most livestreaming platforms today offer detailed analytics that allow you to see data about the engagement of your viewers. Each platform is different in the data that they offer. For example, Facebook Live streams offer broadcasters data about the various engagement emojis viewers used and detailed information about viewer demographics. YouTube offers more information about chat rates and watch time.

THE BASICS OF LIVESTREAMING

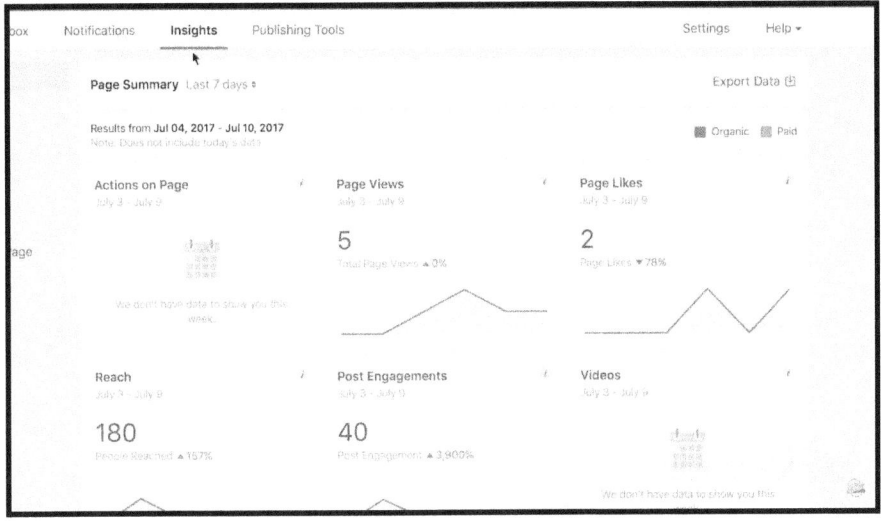

Reviewing Facebook Page Insights.

Overall, engagement is a valuable tool for measuring the success of your livestreams because it means that you are invoking a reaction worth attention in a busy technology-saturated environment. Consider what matters for your business and measure the metrics that really matter. If you are looking for sales, measure sales. If you are looking for new leads, measure that. Engagement may not have a direct impact on your near-term goals, so consider it a real-time metric for actions that may come in the future. Look for anecdotal evidence first, and then look for long-term trends that impact your most important goals.

6 HOW TO LIVESTREAM TO YOUTUBE

How can I livestream on YouTube?

YouTube is arguably the most powerful free livestreaming destination available today. YouTube is both easy to use for beginners and offers advanced capabilities for power users. Those new to YouTube will enjoy easy livestreaming options such as streaming from its smartphone app or launching a livestream directly from a computer's webcam. Those who have livestreamed for years, like YouTube's 4K streaming options, 360 video, DVR, live super chat donation system, and API integrations.

The "Go live" function on top right.

Example of livestreaming to YouTube using a phone.

How can I livestream to YouTube from my computer?

You can livestream to YouTube from your computer in a couple of different ways. In order to livestream to YouTube with your computer, you will need to verify your YouTube account. Once verified, you can enable livestreaming though it may take up to 24 hours to take effect.

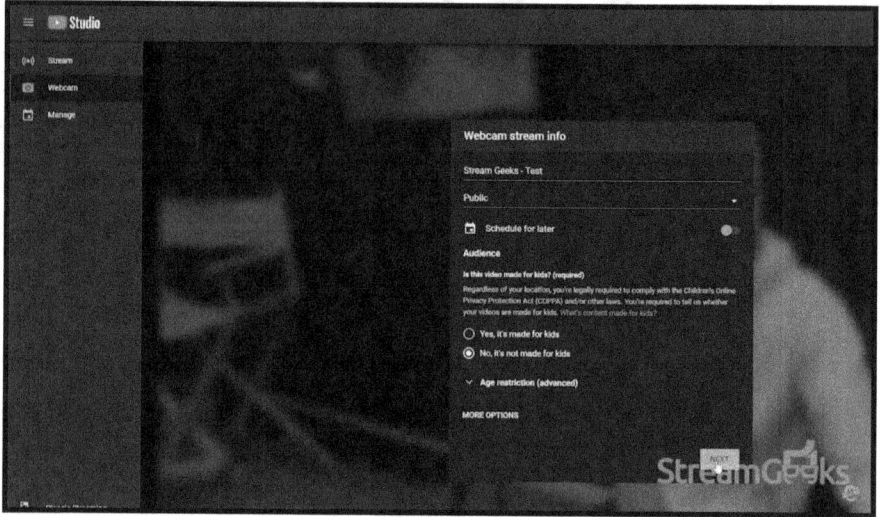

Livestreaming to YouTube with a webcam.

Perhaps the easiest way to livestream to YouTube is directly through the YouTube website with your webcam and microphone. Start by logging into YouTube and clicking the "Go live" button in the top right corner. Here you have the option to schedule a livestream, start a live stream with your webcam, and monitor your live channel. If you choose a webcam, you can start a livestream with any webcam and microphone connected to your computer. When you start, YouTube will create a snapshot from your computer to use as the thumbnail. You can quickly fill out some descriptive details for your livestream and click "Go live" to start your stream.

Should I use Stream Now or schedule my live streams on YouTube?

You have two unique streaming options with YouTube. You can use your Stream Now channel or you can schedule a live stream. When you schedule a livestream, a unique video page and URL are created for your upcoming event. When you use your stream now, the URL always stays the same. Your Stream Now, the URL is YouTube.com/YourChannel/Live. The benefit of scheduled livestreams is that your livestream will automatically become an on-demand video with the same link created before the event. All of the views and view-time your video receives during a livestream will be attributed to your on-demand video. This can be very good for search engine optimization (SEO).

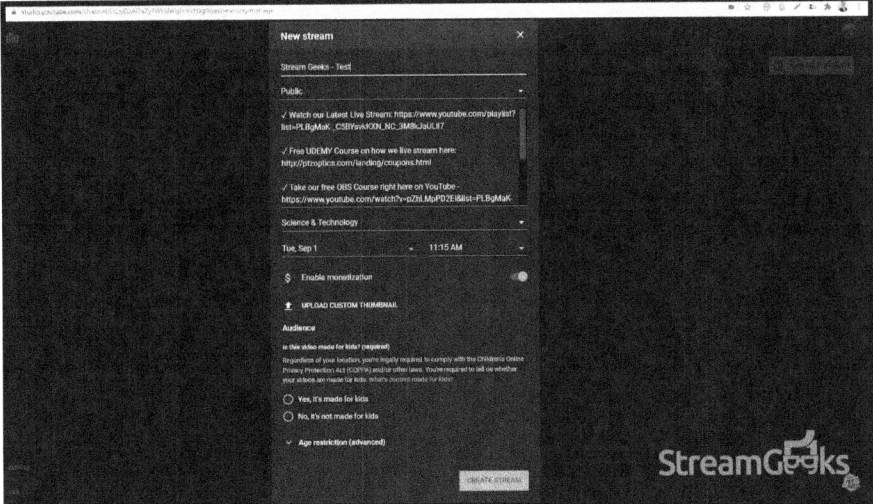

Scheduling a livestream on YouTube.

How can I live stream to YouTube from professional video production software?

It's quite easy to live stream to YouTube with professional video production software. Whether you schedule your live stream or use Stream Now, you will be given a RTMP URL and secret key to enter into your live video production software. This information will allow software like OBS, Wirecast, or vMix to livestream directly to YouTube. If you use Stream Now, your livestream will start as soon as YouTube receives the video. If you schedule your livestream, YouTube will receive the stream to preview, but you have the ability to start the stream with the click of a button.

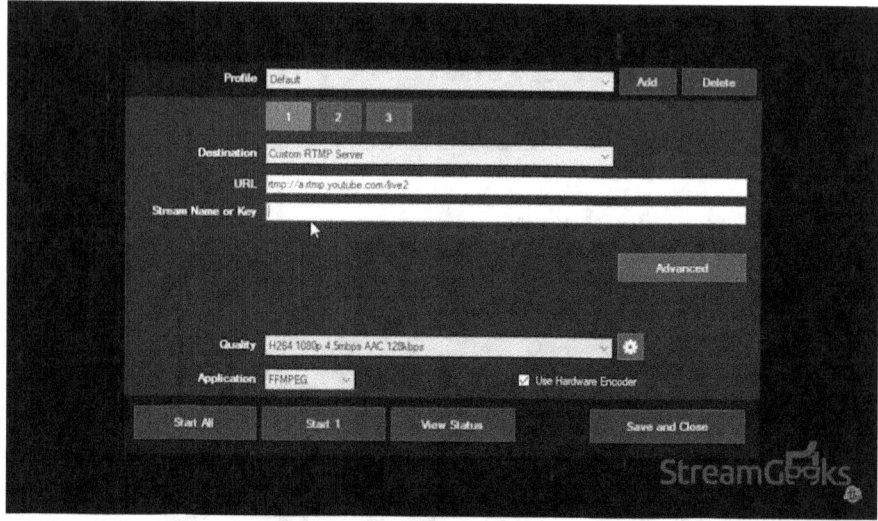

Entering RTMP server information into vMix.

7 HOW TO LIVESTREAM TO FACEBOOK

Did you know that you can livestream on Facebook for free? You can livestream to your personal Facebook page, a group, a business page that you manage, or even inside of a Facebook Messenger Room.

How to livestream from your smartphone?

The easiest way to livestream to Facebook is with the Facebook app on a smartphone. When you first open the Facebook app you will see a "Live" button that you can click. When you click this button it will give Facebook access to your smartphone camera and microphone. Facebook will then give you the option to choose who can see your livestream. You can choose between public, friends, a custom mix of friends, or only you. If you scroll down you will also see options for streaming directly into groups that you are a part of.

Pro Tip: If you enjoy livestreaming with the Facebook app, consider updating your Facebook app frequently. New features such as video calling, allow you to bring a friend into your livestream. You can wear masks, add a donate button, and add all sorts of visual effects to your livestream

THE BASICS OF LIVESTREAMING

How to livestream to Facebook with a smartphone.

How to livestream to Facebook from Messenger Rooms?

Messenger Rooms is a new feature inside of Facebook. These are private rooms that you can use to host video calls with up to 50 other people. You can now livestream these chat rooms by clicking the "Live" button in the upper right corner. Give your live video a title and select where you want your livestream to go. This is a great way to livestream an interview or group meeting.

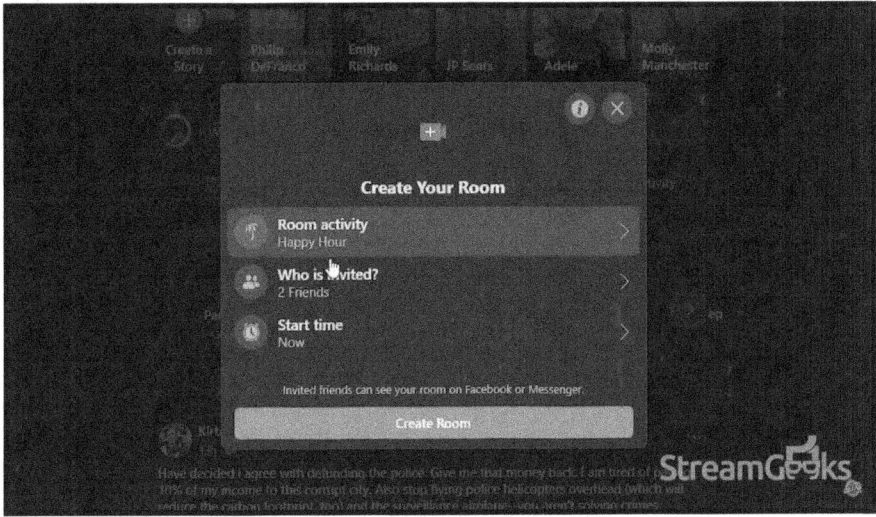

Create a livestream room on Facebook.

How to livestream to Facebook with a webcam?

You can use your computer's webcam to livestream directly to Facebook as well. To enable this, click the "live" button on Facebook which will open up the livestreaming area of Facebook where you can enter a title and description for your livestream. Select the "Use Camera" feature and allow Facebook to access your camera and microphone. This will offer you a live preview of your video and audio to check before you appear live to the audience. Now click the "Go live" button and start your livestream. Before going live, consider creating some polls and questions to drive user engagement.

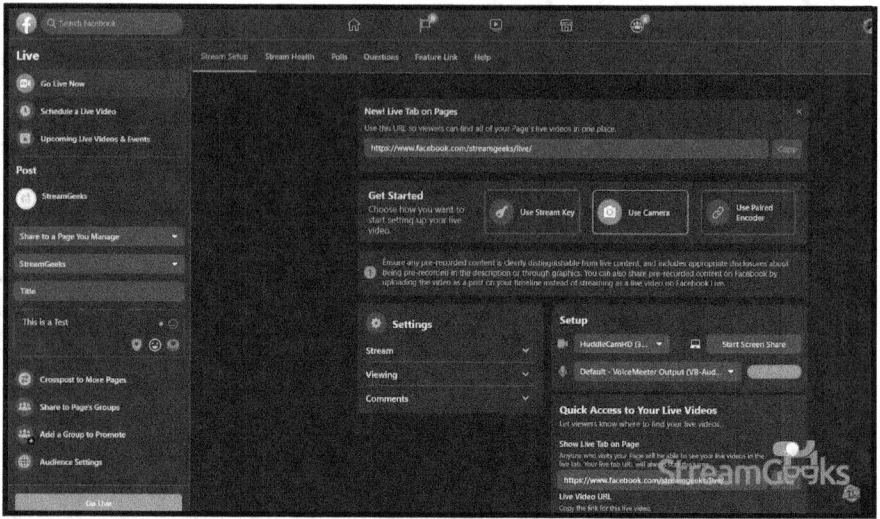

Facebook's livestreaming control area.

Scheduling a Facebook livestream is a great way to notify your audience of an upcoming livestream. When you schedule a livestream, Facebook will notify your friends and followers before the livestream actually starts. Schedule a livestream on a page by clicking the "Live" button and selecting the "Schedule a Live Video" option. When you schedule a livestream, you will choose the time and date that Facebook will set up for your event, and give the livestream a title and description here. Facebook will automatically create a StreamKey that you can use with your encoder. Then copy this stream key information into a streaming encoder such as OBS, Wirecast, or vMix to start streaming to Facebook. Facebook-scheduled livestreams will start exactly at the start time that you have set. A countdown timer will appear to let viewers know when your livestream is scheduled to start.

8 HOW TO LIVESTREAM A ZOOM MEETING

Did you know that you can livestream your Zoom meetings to websites like Facebook, YouTube, and LinkedIn? If you have a professional Zoom Meetings license you can enable livestreaming in your Zoom dashboard to unlock live streaming capabilities. This will turn any meeting into a livestream.

A Zoom meeting that is being livestreamed to Facebook.

How do I enable livestreaming with Zoom?

First visit Zoom.US and log in to your account. Navigate to the Account Management tab, click on "Account Settings," and scroll down to the In Meeting (Advanced) area. Click "Allow livestreaming meetings" and select to enable Facebook, Workplace by Facebook, YouTube, and other custom livestreaming services.

THE BASICS OF LIVESTREAMING

Zoom function that nables livestreaming on Pro accounts.

How do I start livestreaming a Zoom meeting?

Now that livestreaming is enabled on your Zoom license, you will see a "more" button on the bottom control bar. When you click this button, you will have the option to livestream your Zoom meeting on Facebook, Facebook Workplace, YouTube, or custom RTMP destinations.

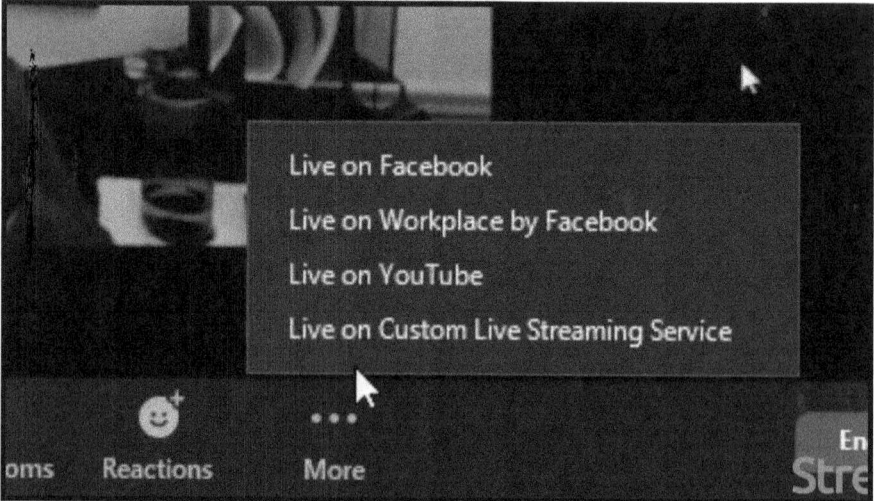

Clicking on "More" takes you to options for livestreaming.

How do I livestream my Zoom meeting to Facebook?

Livestreaming your Zoom meeting to Facebook is easy. First, you will need to have a Facebook account. When you click

31

the "Live on Facebook option" you will have the option to choose where you want to post your live video. For example, you can choose to share it on your timeline, a friend's timeline, in a group, in an event, or page you manage. When you are live, you will see a "Live" button on the top left-hand side of your Zoom meeting. You can stop your livestream using the "more" button and then click the "End Livestream" button.

How do I live stream my Zoom meetings on YouTube?

Livestreaming a Zoom meeting on YouTube is also easy. In order to livestream a Zoom meeting to YouTube, you will need to have a YouTube account with livestreaming enabled. When you click the "Livestream on YouTube" option, Zoom will automatically connect to your YouTube account. Follow the steps to log in to the correct YouTube account and Zoom will start livestreaming to YouTube. When you are live, you will see a "Live" button on the top left hand side of your Zoom meeting. You can stop your live stream by using the "More" button, then click the "End livestream" option.

How do I livestream my Zoom meetings to scheduled livestreams or other destinations like Twitch?

If you have already scheduled a specific event on YouTube or Facebook that you would like to livestream your Zoom Meeting to, use the "Live on Custom LiveStreaming Service" option in Zoom. This allows you to copy and paste the RTMP streaming information from any livestreaming destination for Zoom to livestream to. Inside of your Facebook, YouTube or other livestreaming destination, all you need to do is copy and paste the information into Zoom's custom RTMP streaming area. Once again, when you are live you will see a "Live" button on the top left hand side of your Zoom meeting. You can stop your livestream using the "More" button by clicking the "End livestream' button.

9 WHAT IS OBS?

OBS stands for Open Broadcaster Software, a free, open source live video production software that's supported by a large community of developers from around the world.

What is OBS used for?

OBS can be used for live video production, livestreaming and video recording. When you first download and install OBS, a setup wizard will ask you if you would like to optimize the software for recording or livestreaming. OBS has the ability to mix together many different audio and visual sources into a live video production environment. OBS also supports many plugins which can extend its functionality to include features such as NDI support, VST plugins, and Stream Deck controls.

The StreamGeeks' studio using OBS.

Where can you download OBS?

You can download OBS for free at OBSProject.com. It's available for Windows, Mac and Linux computer systems. At OBSProject.com, you can also read the blog where you will learn great tips for using OBS. OBSProject.com also includes helpful guides in the help section and a user support forum where you can post questions.

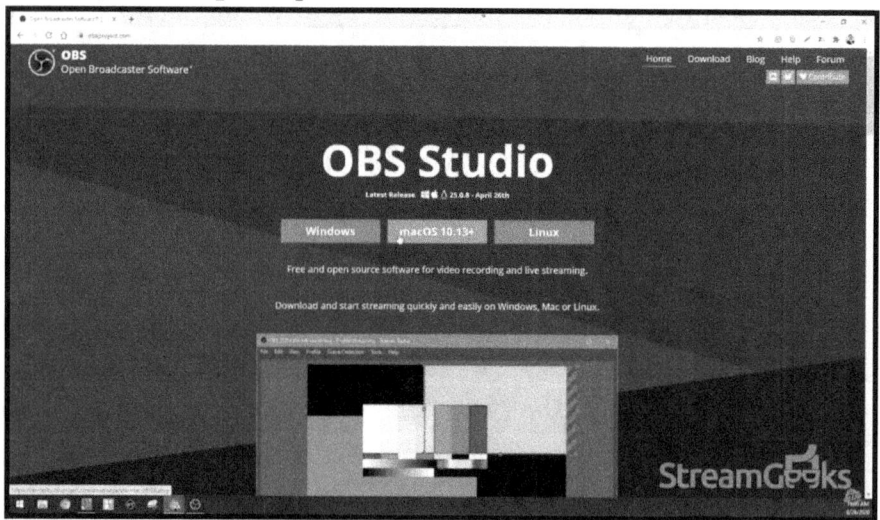

Download OBS at OBSProject.com. **What is an OBS plugin?**

OBS plugins are used to extend the functionality of OBS by adding custom code written to do specific tasks. The most popular OBS plugin adds support for NDI, which is an IP video production protocol. Another popular plugin is called VirtualCam, which allows you to take any video inside of OBS and connect it to another camera via a virtual webcam source. This is great for pumping video from OBS into applications like Zoom. Other popular plugins include a remote control for OBS, which provides an IP server to control OBS remotely, and the PTZOptics OBS plugin which allows you to control PTZ cameras using OBS.

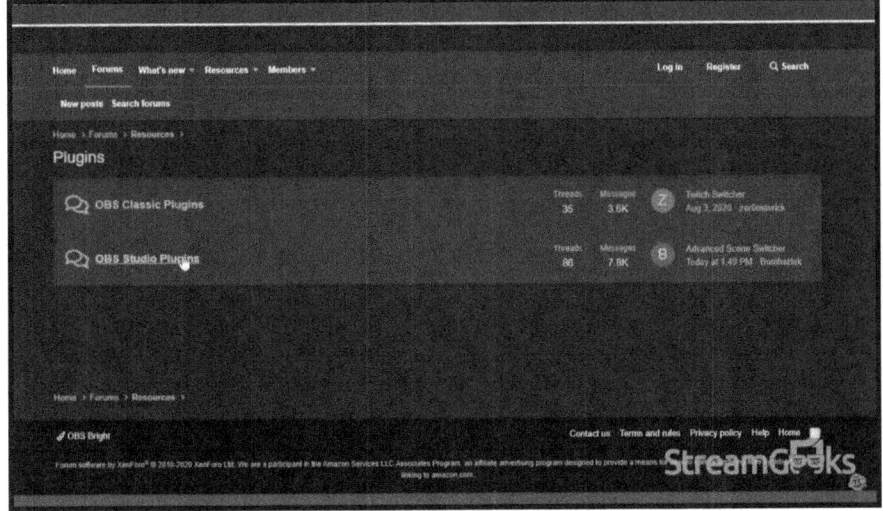

OBS forms where plugins are available to download.

How can I add cameras and microphones to OBS?

OBS organizes all audio and video sources into scenes. In this way, you can create custom scenes with a variety of layouts to display all of your audio and visual sources. Start by selecting a scene to input audio and video sources. Then click the "plus" button in the sources to select a media type. Select the type of audio or video source you would like to add into OBS. Next, give these sources a name. You may be asked to adjust some source properties before adding the

THE BASICS OF LIVESTREAMING

source into your production.

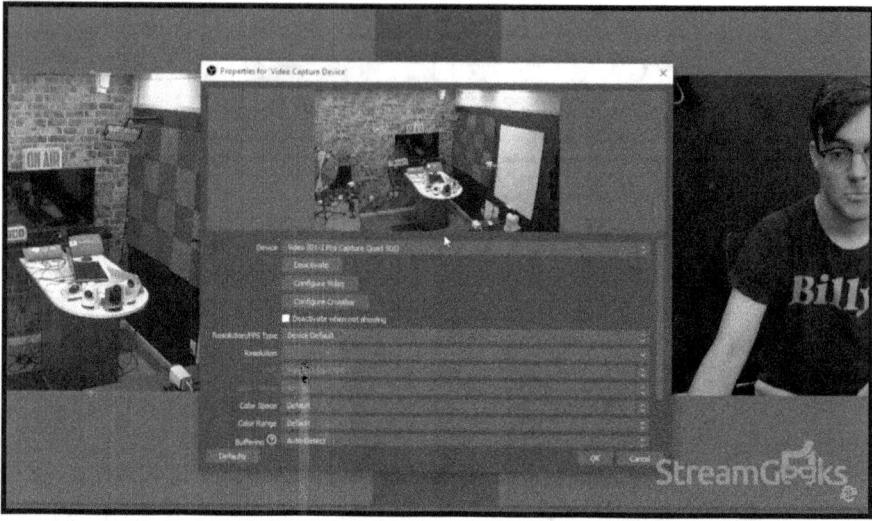

Adding a camera source into OBS.

Additional audio sources can be added to OBS in the settings area under the audio tab.

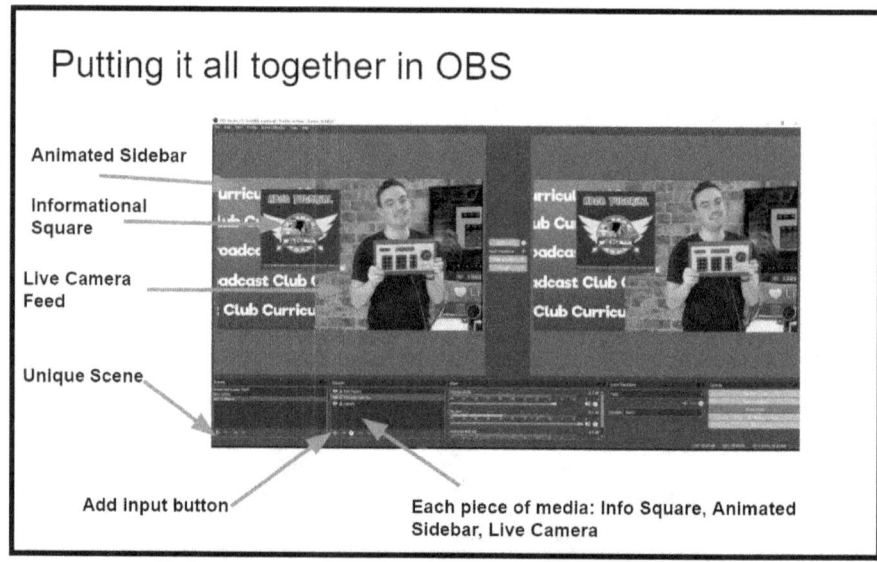

Using OBS to create a professional production.

37

How do I record a video with OBS?

Once your OBS system is set up, record videos with OBS using the "Start Recording" button in the lower right-hand corner. Monitor your video in the main OBS window and audio using the audio mixer levels. Configure the video recording settings and recording location in the settings area inside of the output tab. The video bitrate is used to set the video quality of your recording. By default, this is set to 2500 kbps but you may want to increase this number to 10,000 Kbps in order to increase your video recording quality. Choose a recording format and recording folder to store your videos in this area as well.

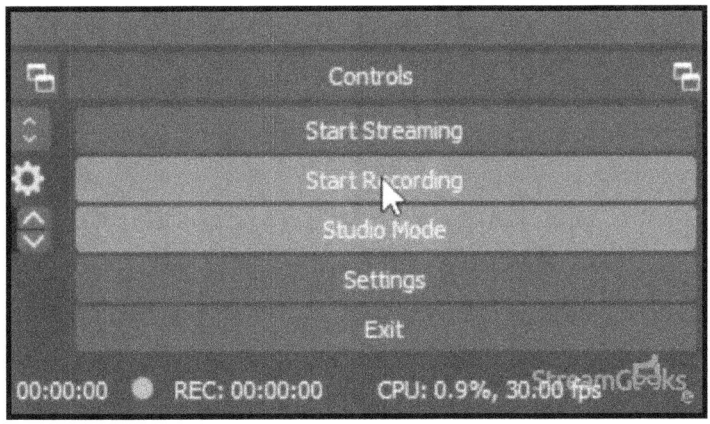

Close-up view of the OBS control area.

How can I livestream with OBS?

Once your OBS system is set up, you can use the "Start Streaming" button to stream video from OBS. You can configure the stream destination in the settings area inside of the Stream tab. Services such as Twitch, YouTube, and

Facebook can all be connected using your accounts' log-in information. You can also click the "Custom" button to enter a custom server and stream key to livestream to. Once you have configured the streaming destination settings, OBS will livestream to this destination when the "Start streaming" button is clicked.

How can I learn more about OBS?

The StreamGeeks offer an *Unofficial Guide to Open Broadcaster Book* via PDF on the StreamGeeks.us homepage. The book is also available in paperback on Amazon.com. Finally, there is an online course that accompanies the book on Udemy. The links appear in the video description below.

10 WHAT IS VMIX?

VMix is live video production software that can turn a regular Windows computer into a professional video production studio. It enables you to mix video and audio sources into a production which can be recorded, streamed and connected to many popular workflows. The output of vMix can be set up in standard definition (SD), high definition (HD), and even 4K. All you need is a PC desktop or laptop with Windows 10 and a DirectX10 compatible graphics card.

Using vMix.

More specific system requirements are available at: https://www.vmix.com/software/supported-hardware.aspx.

The layout of vMix is designed to create the look and feel of a professional broadcast studio in which both preview and output windows appear side-by-side during operation. Some users may feel a bit overwhelmed at first, but will quickly find the interface both intuitive and powerful.

THE BASICS OF LIVESTREAMING

VMix accepts inputs in multiple formats, including cameras, capture devices, NDI (Network Device Integration) sources, video files, and even more advanced input sources such as SRT and RTSP live video streams. VMix then allows users to mix inputs to produce a live production with multiple outputs which can then be used for livestreaming, recording, image magnification, IP workflows and much more.

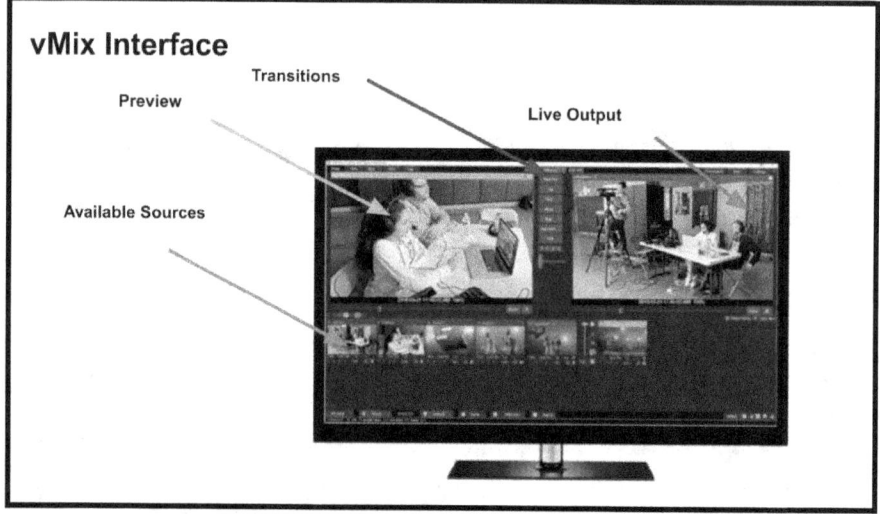

vThe vMix interface is set up like a traditional video switcher with preview on the left and the output on the right.

How Does Vmix Compare to Other Solutions?

There are several video production software options, but vMix hits the sweet spot of features and value for many users and organizations with Windows computers. vMix has a unique pricing strategy that allows users with different budgets and needs to choose the best option. It comes in five editions Basic, Basic HD, HD, 4K, and Pro, ranging in price from free to $1,200.

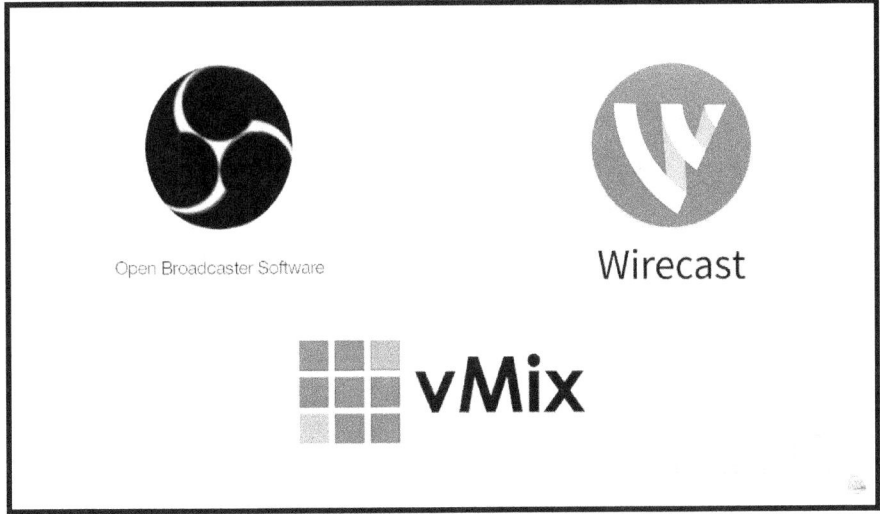

OBS and Wirecast are livestreaming software options.

Many of the best core features are included in even the lowest-priced editions. The least costly paid option, which is $60, offers HD resolution, three camera inputs, overlays, built-in animated titles, scoreboards, and tickers. All versions allow the user to record and send up to three simultaneous livestreams to destinations such as YouTube and Facebook Live. One of the greatest features of vMix is its ability to grow with your production needs. New streamers can purchase a license for what they need at the time and upgrade, adding more features without having to learn a new software environment.

For this reason it's a great choice because
you are unlikely to stop your learning process and switch to different software because of a technical limitation. VMix is used for simple and advanced productions everyday around the world.

What are the best features of vMix?

A few key features that make vMix unique include superior support for NDI. The auto-discovery mode for NDI sources on the network is intuitive and reliable. The output sections of vMix, enable you to configure up to four NDI outputs with advanced features you are unlikely to find in other solutions. The overlay and multi-view features make it easy to mix multiple sources into a production that matches what you see on TV. The built-in animated titles make displaying information a breeze but more importantly, the ability to get dynamic content into vMix is very impressive. You can use vMix's vMix Social Tool which integrates titles with social media sources such as Facebook, Twitter, and YouTube and also pull in Data Sources. The Data Sources function allows you to map data into titles via RSS, XML, JSON, Text Files, and even Google Sheets. VMix Call is a helpful feature that allows you to bring video callers directly into vMix to host talk shows and interviews. Forget about using mix-minus audio mixing, or complicated virtual cables in and out of Skype or Zoom. Using vMix Call is little more effort than sending your guest a link to your production.

	BASIC	BASIC HD	HD	4K	PRO
	FREE	$60 USD	$350 USD	$700 USD	$1200 USD
Total Inputs	4	4	1000	1000	1000
Camera / NDI Inputs	2	3	1000	1000	1000
Maximum Resolution	768 x 576	1920 x 1080	1920 x 1080	4096 x 2160	4096 x 2160
Overlay Channels	1	1	4	4	4
Recording	✓	✓	✓	✓ 2 Recorders	✓ 2 Recorders
Streaming Including 3 simultaneous live streams	✓	✓	✓	✓	✓
Fullscreen Output	✓	✓	✓	✓	✓
External Output	✓	✓	✓	✓	✓
GT Designer Standard 100+ Built-In Animated Titles, ScoreBoards, Tickers	✓	✓	✓	✓	✓
GT Designer Advanced Custom Animated Titles and Import PSD's	✗	✗	✗	✓	✓
Playlist	✓	✓	✓	✓	✓
Professional Colour Correction	✓	✓	✓	✓	✓

A software comparison table.

For all of these reasons plus advanced color corrections that our show could not live without, users rave about vMix. Check it out and give the free 60-day trial a go. And if you want to get started off on the right foot, download our free *Unofficial Guide to vMix* today, and jumpstart your next livestreaming project with the power of vMix. This book is also available via a paperback version on Amazon.

11 WHAT IS THE BEST CAMERA FOR LIVE STREAMING?

When evaluating the equipment needed for livestreaming, the first item to look for should be a camera. While you also may need a microphone, software, and other encoding hardware, considering the type of camera you need for livestreaming is the most important step in building a live video production system.

What are the different types of livestreaming cameras?

Here are all the livestreaming camera types available today, for all budget types.

1. Webcams
2. Camcorders
3. DSLR Cameras
4. PTZ Cameras
5. Broadcast Cameras

What is the best webcam for livestreaming?

Webcams are the most affordable and they offer a variety of features for livestreaming. Most webcams today are HD, which is 1280x720p or Full HD, which is 1920x1080p. Most webcams also support 30 frames per second but some also support 60 frames per second. When you are considering the resolution and frame rate of a webcam for livestreaming, think about the production software you plan to use and the bitrate you plan to stream at. If you plan to stream in 720p resolution at 30 frames per second, most streaming destinations recommend a bit-rate of 2-4 megabits per second. If you plan to stream at 1080p in 30 frames per second, use a higher bitrate between 4-6 megabits per second.

PTZOptics 1080p webcam.

Once the resolution for your livestreaming project has been determined, you can look for additional features such as electronic pan, tilt, and zoom. For example, the HuddleCamHD Pro features a 4K image sensor which digitally zooms in, pans, and tilts just like a PTZ camera. You may also consider using the HuddleCamHD Pro IP which is an NDI camera. But, before we dig into NDI cameras that can be used as webcams, let's cover camcorders.

The HuddleCamHD Pro IP is an NDI webcam.

By the way, if you are still watching this video, hit the "like" button. We check out the likes on all of our videos to see which type of videos you enjoy the most.

What is the best camcorder for livestreaming?

Camcorders will provide great video quality for your next livestreaming project and they are not that expensive. A brand new Canon VIXIA camcorder starts at only $299 and is ideal for zooming in long distances. If you are considering a camcorder for livestreaming, you need to think about how you will connect the camera to your live streaming system. Unlike a webcam, most camcorders do not have a USB port. Therefore, you should look for an HDMI output that you can use with an HDMI capture card. The cardconverts HDMI into a usable USB connection with any computer. Once you connect the USB from the capture card to your computer, you can bring the camcorder into a software like OBS, Wirecast, or vMix just like a webcam.

Standard camcorder.

What is the best DSLR camera for livestreaming?

Many people \ like to use DSLR cameras for livestreaming because they provide great value and performance. DSLR cameras offer isuper sharp auto-focus and interchangeable lenses that can provide beautiful, blurry backgrounds. DSLR cameras have amazing quality but they do come with their own set of challenges when used for livestreaming. Because of the popularity of livestreaming, many DSLR camera manufacturers are adding new firmware and features specifically designed for live-streaming.

THE BASICS OF LIVESTREAMING

A DSLR camera.

Although DSLR cameras were designed for photography and filmmaking, companies like Canon have added features that allow the USB port to be used for live streaming. When selecting a DSLR camera for livestreaming, look for a "clean" HDMI feed which is used with a capture card that does not include the on-screen display menu options. While DSLR cameras require a battery, when they are being used to livestream, they are typically plugged in for a long time. In the early years of DSLR cameras being used for livestreaming, prolonged use led to overheating. Today, most manufacturers have adapted their models to address this issue.

What is the best PTZ camera for livestreaming?

PTZ cameras are ideal for livestreaming because they combine the ease of use of a webcam with the functionality of a camcorder. This is because pan, tilt, and zoom cameras almost always include optical zoom which is used to zoom into subjects from long distances. What makes PTZ cameras unique is their ability to be remotely controlled. For example, PTZOptics cameras can be controlled with software solutions such as vMix, Wirecast, OBS, Livestream Studio, and Mimolive. This allows a one-man production to operate the livestreaming software and automate camera controls for one or more cameras.

A PTZ camera.

PTZ cameras are built for 24/7 use and offer easy installation options such as Power Over Ethernet so you can power a PTZ camera using a single ethernet cable from your network. PTZ cameras are also very small and discreet. This makes them ideal for installing ina church, or any space that you are adding livestreaming. PTZ cameras can be installed on walls, ceilings, and even under balconies.

What is the best broadcast camera for livestreaming?

Broadcast cameras are used for professional video production environments and cinema. Cameras such as the Blackmagic URSA or Sony over-the-shoulder style cameras are expensive, but they offer large image sensors and unmatched quality. If you are considering a project with professional broadcast cameras, it's nice to know that these cameras can also be used for field shooting and live work. If you are installing multiple broadcast cameras in the same space, it's important to test them in your studio environment. Many professional projects will use the same camera make and model to ensure consistent color matching and quality throughout the project. Consider professional broadcast cameras from Blackmagic, Sony, and Panasonic.

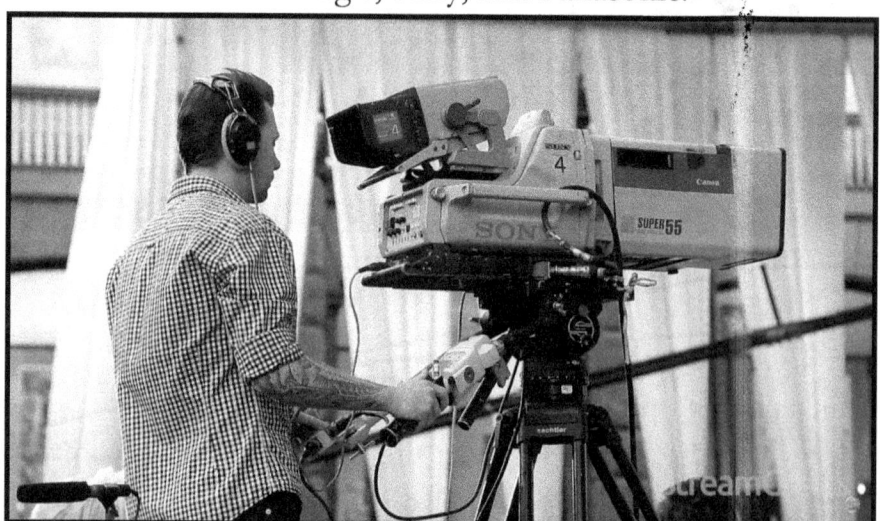

A camera operator uses a professional broadcast camera.

Most professional broadcast cameras connect to live video production systems via SDI but there are high quality wireless connections available as well. These SDI video connections feature locking connectors that are ideal for high-profile applications such as sports, television and production. Examples of SDI-based broadcast systems include Grass Valley video switchers, NewTek Tricasters, and Roland video switchers.

What is the best NDI camera for livestreaming?

Some would argue that the best NDI camera for live streaming is your smartphone. For example, with the latest NDI HX apps available for iOS, you can send 4K high quality video over WiFi into your video production switcher such as OBS, Wirecast, vMix, Livestream Studio, and Tricasters. NDI cameras are also available in the form factor of a webcam, a PTZ camera, and broadcast-style over-the-shoulder cameras. NDI is a technology that in many cases can replace SDI because of its easy implementation.

NDI-enabled PTZ cameras.

So what is the best camera for livestreaming?

The best camera for livestreaming is the camera that you have. The quality of a great webcam just might surprise you once you adjust the lock in the focus and adjust the color settings, like on the PTZOptics PT-WEBCAM-80, for example. The quality of your DSLR will improve plus you can use it to shoot great pictures for your next project. If you are permanently installing cameras in a space or setting up for larger venues, PTZ cameras will be your best bet. It's all about understanding your application of livestreaming and perhaps one day, you will be shopping for broadcast cameras to shoot a Hollywood-style livestream.

9 HOW MUCH BANDWIDTH DO I NEED TO LIVE STREAM?

Bandwidth is a term used to describe the amount of data transfer speed available between two locations. For example, if your internet service provider offers you 1 gigabit of download and upload speeds, you would have 1,000 megabits of bandwidth for livestreaming. When livestreaming, you're sending video somewhere that will require certain upload speeds.

Because bits are such a small unit of data, bandwidth speeds are usually referred to with the prefixes like kilo, mega, or giga. Kilo means thousand, mega means million, and giga means billion. So, for example, 1 kilobit is the same as 1 thousand bits, 1 megabit is 1 million bits, and 1 gigabit is 1 billion bits.

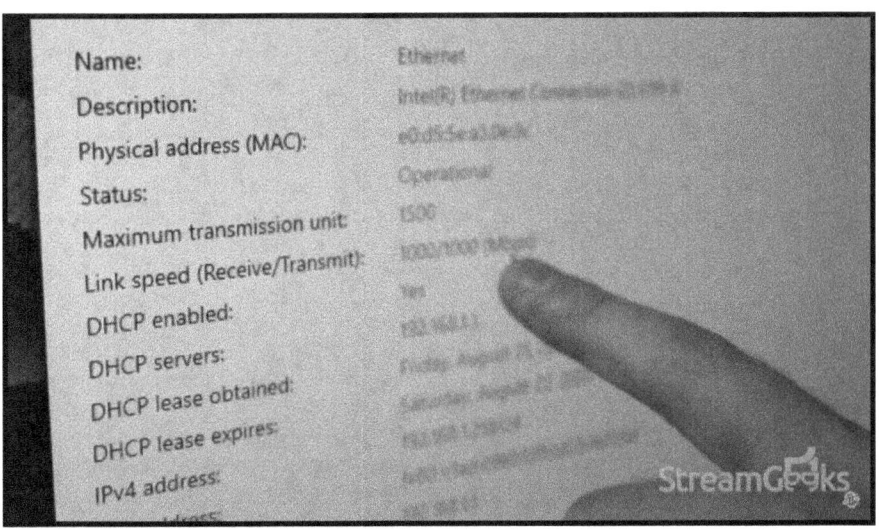

Ethernet connection speed on a Windows computer.

THE BASICS OF LIVESTREAMING

These same prefixes also apply to bytes. It is important to understand however, that bits and bytes are not the same. A byte is a measurement of data storage. A bit is the smallest unit of computer information, representing either a 1 or a 0. Data transfer speeds are represented in bits per second, and data storage is measured in bytes.

When you talk about bandwidth, there are two different processes taking place; upload and download. Download is the process of retrieving data from the internet, while upload is the process of sending data from your devices to the internet. This is the same process that happens when downloading a file from the internet, however within the context of livestreaming, its application differs.

The main difference between downloading and uploading files as opposed to streams, is that files usually do not require a minimum bandwidth to transfer data properly. File downloads and uploads will simply operate at whatever speed your network's bandwidth allows. Streaming, on the other hand, depends on the production setup. When you start livestreaming, you generally have to choose a bitrate which encodes your audio and video into a stream of information that uses upload bandwidth. It is essential to know your network's upload and download speed because these rates can affect the quality of your outgoing streams. You can check your bandwidth speeds by going to Google and searching for "Speed Test".

THE BASICS OF LIVESTREAMING

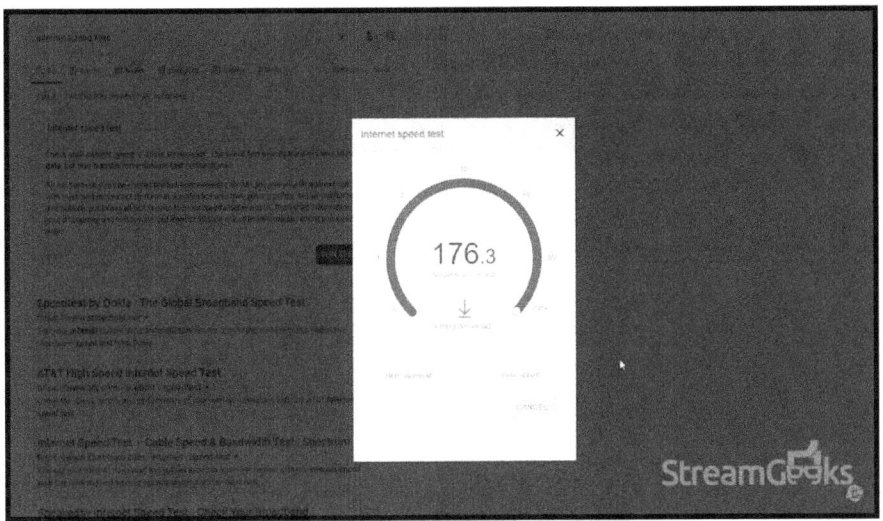

A bandwidth speed test.

To determine how much bandwidth you need to livestream, evaluate the CDN's recommended settings. Facebook, YouTube, and Twitch are all CDNs that provide RTMP server information to stream with. RTMP, (Real Time Messaging Protocol) information, is available as a server name and a secret key. Inside Open Broadcaster Software (OBS), for example, in the Stream area, you can enter your RTMP information. This is the destination for your live stream and you will need to decide the bitrate livestreaming. Search for recommended streaming settings for your desired CDN. For example, Facebook recommends a 4000 Kbps (4 Mbps) bitrate at 720p at 30 frames per second.

Resolution	Pixel Count	Frame Rate	Quality	Bitrate
4K 30fps	3840x2160	30fps	High	30Mbps
4K 30fps	3840x2160	30fps	Medium	20Mbps
4K 30fps	3840x2160	30fps	Low	10Mbps

1080p60fps	1920x1080	60fps	High	12Mbps
1080p60fps	1920x1080	60fps	Medium	9Mbps
1080p60fps	1920x1080	60fps	Low	6Mbps
1080p30fps	1920x1080	30fps	High	6Mbps
1080p30fps	1920x1080	30fps	Medium	4.5Mbps
1080p30fps	1920x1080	30fps	Low	3Mbps
720p30fps	1280x720	30fps	High	3.5Mbps
720p30fps	1280x720	30fps	Medium	2.5Mbps
720p30fps	1280x720	30fps	Low	1.5Mbps

The bitrate you select is the bandwidth you will use for uploads.

Once you know the recommended bitrate to livestream, run an internet speed test. Search for "Speed Test" in Google and you will instantly receive a report on your computer's internet access speed.

You can think about your livestream's resolution as the size of your overall canvas. The bitrate you select is the amount of data used to fill that canvas. You could have a high-quality 1080p stream with a bit rate of 6 Mbps or a low-quality 1080p stream with a bit rate of just 2 Mbps. Years ago, back in the time of SD (320×240 pixels), you could use flash to encode and stream at roughly 500 Kbps (that's half a Megabit). But today, most audiences will expect a minimum of 720p video and a bit rate of at least 1.5 Mbps.

Pro Tip: If you have limited bandwidth, don't stretch your canvas with a low bitrate. Try using 720p instead of 1080p and the video quality will actually look better at low bitrates.

Under certain circumstances, you may need to choose between livestreaming a single high-quality video stream or multiple live streams of lesser quality. For example, if you have 10 Mbps of upload speed, you may create a 3 Mbps stream to YouTube and a 2Mbps stream to Facebook, while maintaining a healthy 50% of bandwidth headroom. If you are concerned about creating a single high-quality stream, stream to YouTube using 5Mbps.

Keep in mind that you can always record high-quality recording to your local hard drive. Many production experts record a "high bitrate" MP4 file ranging from 12-100 Mbps. The recordings saved to your local hard drive will always be higher quality than the live streamed recordings available on YouTube and Facebook. The higher the bitrate you use, the larger your file size will become. I generally use between 8-16 Mbps for my standard video recordings.

Note: Always have some extra upload bandwidth available.

This extra "headroom" acts as a buffer to account for any fluctuations in your network and will result in a more reliable stream.

In order to take full advantage of your network bandwidth, you will need hardware with matching or superior bandwidth capabilities. This means buying a motherboard and/or switcher with an ethernet port that is rated for your bandwidth speed. You will also need to make sure your ethernet cable is rated for these speeds to get full access to your available bandwidth. Ethernet cables are usually rated based on their bandwidth capabilities. For example, CAT4 cabling is capable of providing 16 megabits per second, while CAT5 cabling can reach 100 megabits per second. CAT5e and CAT6 are rated for up to 1000 megabits per second, and CAT7 is rated for 10 gigabit applications.

Cable Name	Bandwidth	Maximum Distance
Cat 5e	1 Gbps	328' (100 meters)
Cat 6	1 Gbps	328' (100 meters)
Cat 6a	10 Gbps	328' (100 meters)
Cat 7	10 Gbps	328' (100 meters)
HDMI 1.4	10.2 Gbps	50' (15 meters)
HDMI 2.0	18 Gbps	50' (15 meters)
SDI	270 Mbps	1000' (300 meters)
HD-SDI	1.5 Gbps	300' (90 meters)
3G-SDI	3 Gbps	200' (60 meters)
USB 2	480 Mbps	15' (5 Meters)
USB 3.0	4.8 Mbps	9' (3 Meters)
Thunderbolt	30 Gbps	3' (1 Meter)

Maximum Distance refers to the longest length the cable can be reliably extended in feet or meter measurements.

Pro Tip: Software like OBS, has an option to use a dynamic bitrate. This allows your upload stream of information to dynamically change based on the available bandwidth with your internet connection. Using this feature can reduce the amount of dropped frames you get during your broadcast.

13 WHAT TYPE OF COMPUTER DO I NEED TO LIVE STREAM?

Almost any computer can be used to livestream today. With that being said, every computer has its limitations. IIf you use OBS, Wirecast, or vMix software on your computer, start by consulting the minimum recommended specifications for the software.

For example, Wirecast recommends an i5 computer processor @ 2.5 Ghz for streaming in 720p, and an i7 computer processor @ 3.0 Ghz for 1080p streaming. Each of these specifications also requires a minimum of 4GB of ram and 2 gigabytes of free hard drive space.

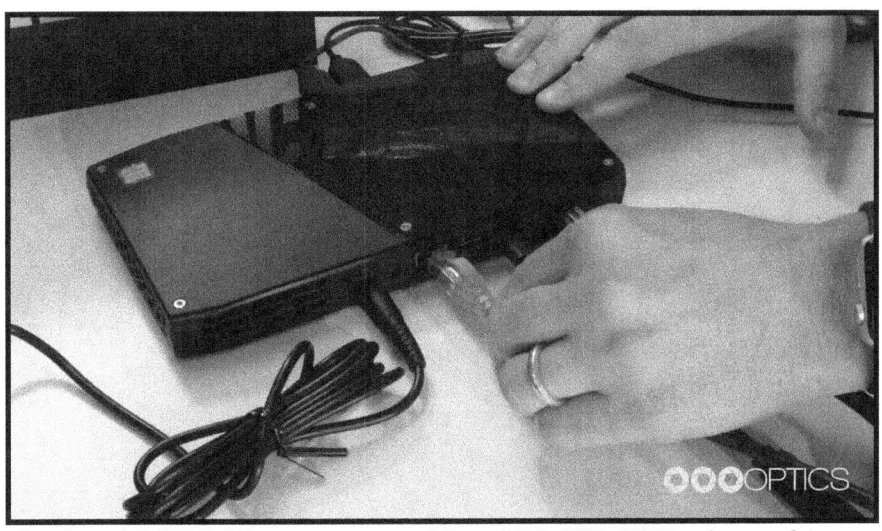

An Intel NUC i7 computer with a 256 GB SSD plugged into an ethernet LAN.

In general, you should consider using an i7 processor with 16 GBs of RAM and a solid state hard-drive for any new live streaming project with more than one or two cameras. AMD often offers a better price to performance ratio than Intel and itoffers significantly better processors for livestreaming than any other company.

CPU core count/thread count as well as clock speed are of equal importance for livestreaming. So purchasing a CPU with the most cores/threads and highest clock speed within your budget is your best bet to achieve optimal performance.

Note: Different CPUs have different thread to core ratios. It is important to compare the thread count of CPUs that have the same amount of cores. For example, one 8 core CPU may use 8 threads, while another different 8 core CPU may use 16 threads. More threads is almost always better when comparing CPUs with the same number of cores. So the 16-thread 8-core CPU will likely outperform the 8-thread 8- core CPU. Most CPUs do not use more than 2 threads per core, however many use only 1 per core instead of 2.

Livestreaming computer on-site running vMix.

Having more and faster cores results in higher multi-thread counts as well as faster thread processing, thus more instructions can be executed in the same period of time. This reduces overall processing time and latency, while also improving multitasking ability so that more programs and cameras can be run simultaneously. These capabilities are crucial for live- streaming, videogaming, and large productions. Ultra high- end streaming solutions that are used by large organizations like ESPN, may use up to 128 or even 256 cores for extremely high resolutions and low latency. On the other hand, low resolution, high latency streams that are used for applications like security camera systems may only require 2-4 cores. Simply put, the higher the stream quality and complexity, the more cores/clock speed you will need.

The 4-16 core range of CPUs offers the best streaming performance relative to the price for the average livestream. Where your use case falls on this spectrum is determined by the amount of multitasking you plan on doing during your livestreams as well as the stream quality you want to achieve. If you are streaming with only 1 program open and 1 or 2 cameras at 1080p, 4-6 cores is all you need. If you plan on having multiple programs open at once and/or streaming with multiple cameras, you will likely need 6-10 cores to achieve smooth performance. Ten-16 core CPUs are only necessary for semi-professional productions using more than 6 cameras which often use CPU-intensive programs simultaneously, such as video editing software or 3D/CGI programs.

Another important component of most livestreaming computers is the graphics card. For example, vMix recommends an NVIDIA GTX 1660 for livestreaming systems using 4 1080p cameras and an NVIDIA 2080 Ti graphics card for up to 6 4K cameras. Graphics cards are essential to most livestreaming software solutions because they take the processing off your main CPU and handle it in the graphics card. This helps reduce the number of dropped frames for video game streamers and rendering time for 3d programs during a stream.

Choosing a GPU is very similar to choosing a CPU in the sense that higher clock speeds are better. Where they differ is, instead of focusing on core count for GPUs, we will focus

THE BASICS OF LIVESTREAMING

on VRAM, which is the memory the graphics card uses to hold information processed by the CPU. The absolute bare minimum recommended VRAM for livestreaming is 2gb. For many applications this is not realistic, however there are use cases where it is all that is needed. For the vast majority of streaming applications, 4-12 gb of VRAM is sufficient. Just like with CPUs, you will need to account for the level of multitasking you do during your livestream and choose a GPU with sufficient VRAM for your purpose. Typically, 4-8gb cards are sufficient for up to 4 1080p cameras, while 8-12 gb cards are capable of handling up to 6 4k cameras. If you want to use even more 4k cameras in your stream and/or game while using 4-6 4k cameras, you will likely need a card with 12 - 24 gb of VRAM.

Pro Tip: Make sure to go into your livestreaming software and ensure that you have enabled it to use your graphics card. Even some simple Intel NUC and laptop computers have integrated Intel Graphics cards that can be used for basic live video production systems.

The same logic can be applied to choosing DRAM, which is the computer's system memory, as opposed to the dedicated video memory found in a GPU. Like VRAM, clock speed and memory quantity are the most important factors for streaming. Where DRAM differs from VRAM, is the amount of it that is required for a smooth system performance. The simplest streaming set-ups using 1-4 cameras at 1080p will require 8-16 gb of DRAM, while using more cameras at

1080p or multiple 4k cameras will require around 16-32 gb of DRAM. More intensive multitasking situations, like Gaming in 4k while streaming multiple 4k cameras simultaneously, or Using VRAM intensive programs while streaming, such as video editing or CGI software, can require 32-64gb of VRAM for optimal performance.

Note: All modern computer platforms (since 2017) have switched to DDR4 DRAM compatibility only, and most Intel platforms have shifted away from DDR3. So if you're looking at an 8th or 9th Gen Intel CPU or later, or an AMD Ryzen processor, you'll need DDR4.

Custom computer by ThinkComputers.org.

The type of storage you choose for your livestreaming PC will have very little effect on your livestream performance. In general, SSDs are better in almost every way than HDDs, however buying a HDD instead of an SSD can be a good way to save money on a streaming computer without losing streaming performance. However, using HDD vs. SSD will affect the performance of some non-streaming tasks, so a SSD is always recommended over a HDD if your budget allows for it. In addition, SSDs are also significantly quieter than HDDs as they have no moving parts, which is a benefit for live productions requiring quiet on set.

Assuming that you understand the processor and graphics card requirements for your next livestreaming computer, you also need to think about inputs and outputs. For inputs, you have a bunch of options. Start by counting the number of USB ports you think you will need. Obviously you'll need a keyboard and mouse, but what about your audio mixer and a secondary USB controller like the Elgato StreamDeck? You may also want to get a built-in HDMI or SDI input. You can insert PCIe cards into many desktop computers in order to create multiple SDI or HDMI video inputs. In most cases, if you only need two cameras, it's easiest to purchase two HDMI or SDI to USB capture cards. However, if you need 3 or 4 cameras, it's easier and more affordable to purchase and install a PCIe card. You can purchase PCIe cards that can be configured to provide a couple of inputs and extra video outputs for your project. Keep in mind that most graphics cards also provide unique outputs that power confidence monitors, multiview monitors, and more.

Finally, you can use additional video switching hardware to take the processing load off of your computer and handle it in hardware. For example, if you need to set up a 6-camera livestreaming system, you can do most of the video processing and switching using a Blackmagic ATEM switcher. Then you can use a simple computer to capture a single mixed output with a capture card into a software like OBS. This approach is effective because hardware switchers rarely have computer issues such as Windows updates. The ATEM Mini is a perfect example of an affordable 4-input switcher with a built-in USB output for streaming or recording. The trade-off with hardware systems is the lack of flexibility that software switchers provide given their access to multiple cameras.

Now you know enough to be dangerous and purchase a very powerful livestreaming computer. The good news is that you won't have to spend nearly the kind of money people used to on livestreaming hardware.

14 WHAT TYPE OF CABLES ARE USED FOR LIVE STREAMING?

Livestreaming from a smartphone may require no cables at all; just charge up your phone and make sure you have a strong WiFi connection.

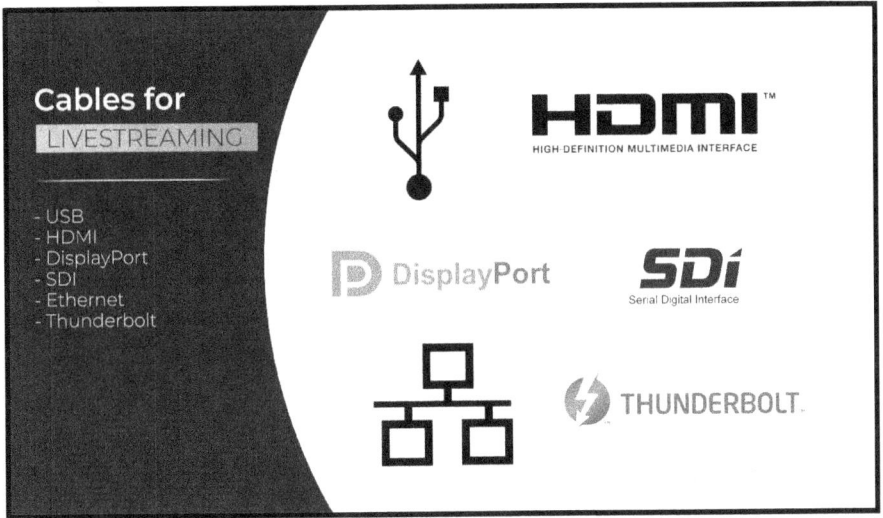

Commonly used video cables.

But many livestreaming systems feature a variety of cables that you should be familiar with, including **USB, HDMI, DisplayPort, SDI, Ethernet, Thunderbolt, XLR, audio cables,** and more.

USB CABLES

Perhaps the most common cable used in any modern video project is the USB cable. USB cables are often used to connect simple devices such as keyboards and mice to your computer. But they are also used to capture video sources such as HDMI or SDI and convert them into usable USB connection sources for your computer.

USB cables come in a variety of connector types including **USB-A, USB-B, USB-C, micro-USB, and mini-USB.** USB cables are currently available in three main versions **USB 2.0, USB 3.0, and USB C.**

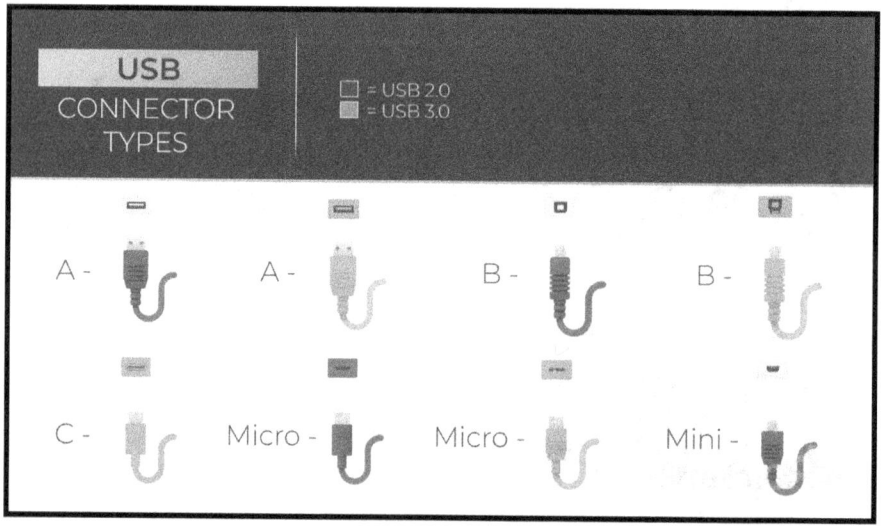

USB connector types.

In general, USB 2.0 cables are used for low bandwidth devices such as audio mixers, keyboards, and webcams. USB 3.0 and USB C cables are used for higher bandwidth connections and with capture cards and professional cameras.

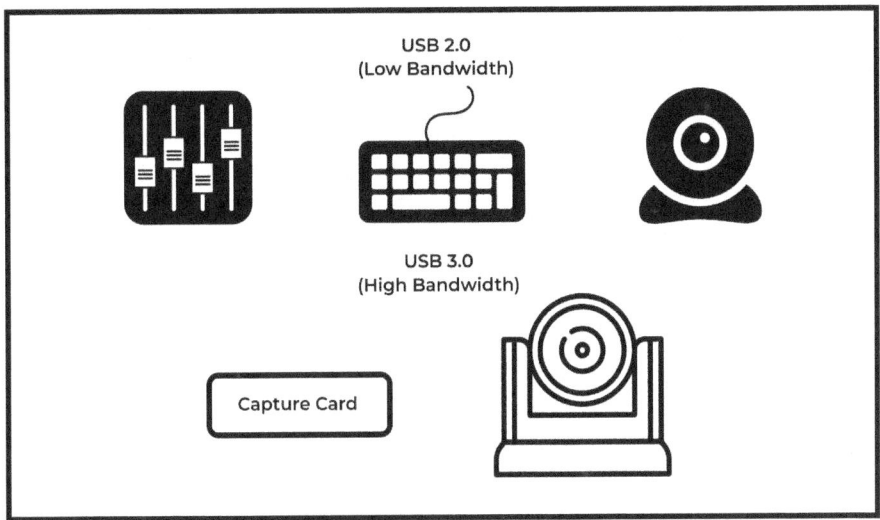

Low bandwidth vs. high bandwidth USB devices.

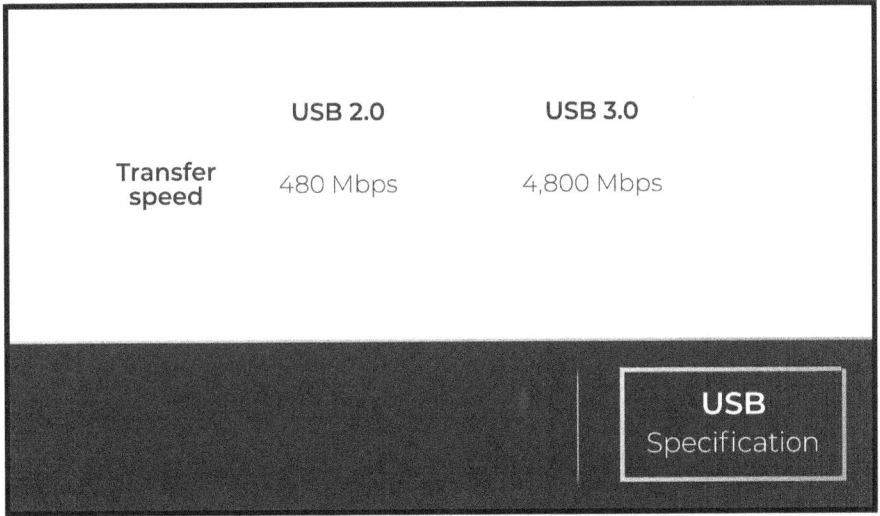

USB bandwidth specifications.

When using USB cables in your video production system it's important to think about cable lengths, the availability of ports on your computer, and USB bandwidth. If you need to extend the length of a USB cable, you can do so with a USB extender.

Pro Tip: If you have to extend a USB connection to a camera more than 20 feet, consider extending HDMI or SDI instead. SDI cabling, for example, can be reliably extended hundreds of feet. It's much more reliable than USB extensions that claim they can be extended this far.

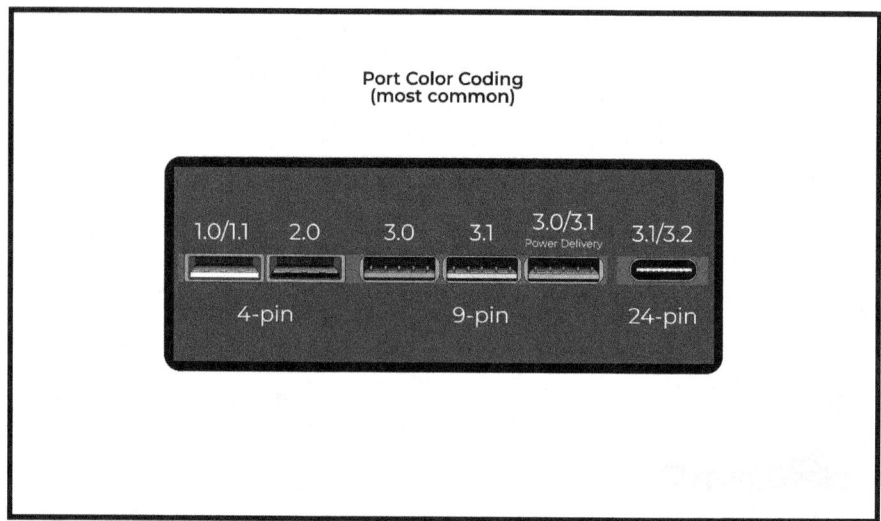

Color-coded USB ports.

USB CABLE PORTS

Next you need to think about USB ports on your computer. Some of your ports may be USB 2.0-capable and others may be USB 3.0-capable. Additionally, these ports may share a common USB bus which could limit the overall bandwidth available to all USB ports in that section of your computer's hardware.

Pro Tip: If you are trying to connect two USB cameras to your computer and notice issues, try switching USB ports. Sometimes two USB ports on one side of a laptop share a USB bus.

By taking one of the ports and switching it to the other side of the laptop, you gain access to additional bandwidth with an extra USB bus internal to your computer.

Be aware of HDMI, SDI, and DisplayPort video connection cables. HDMI and DisplayPort cables are great for short cable runs and they're ideal for connecting LCD monitors and cameras that are within a short distance of each other.

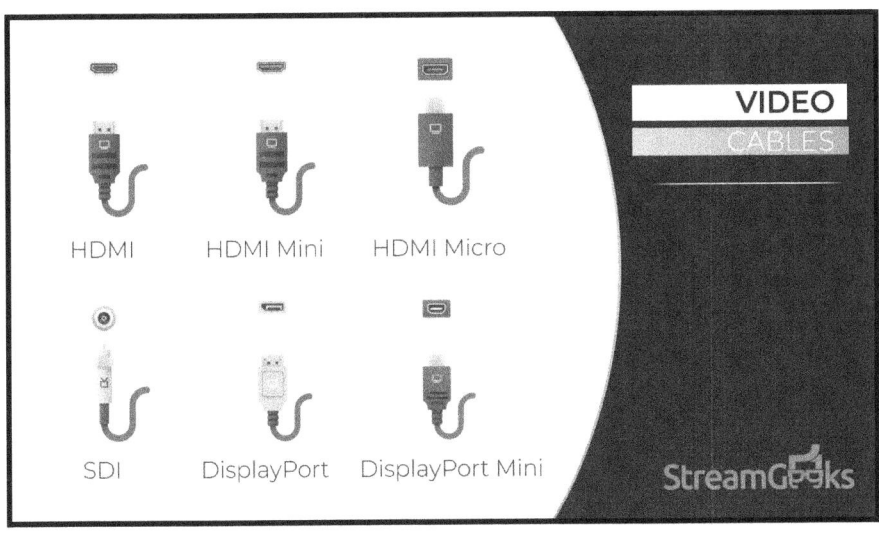

Types of cables used for livestreaming projects.

If you need to extend HDMI cabling, purchase an HDMI extender. But know that HDMI cables should not exceed 50 feet. SDIs cable can be used at much longer distances and have locking connectors, which help to avoid accidental disconnections. For this reason, many professionals use SDI cabling paired with an SDI to USB converter to extend the USB cable for video purposes.

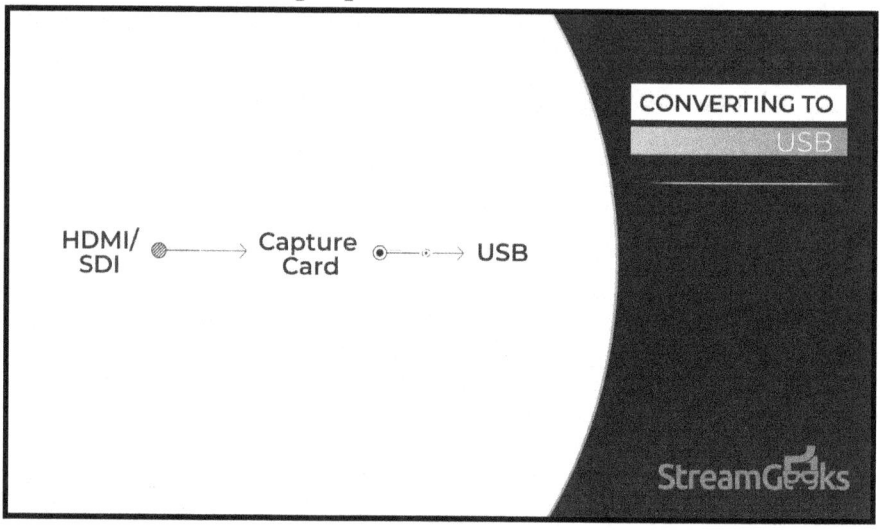

Diagram shows HDMI or SDI cabling with a capture card that converts video sources into USB.

Ethernet cables

Ethernet cables are the heart of many IP-based video production systems and they can also be used to provide your computer with internet access. Ethernet cables rarely extend beyond 328 feet, though they come in a variety of quality types.

THE BASICS OF LIVESTREAMING

Ethernet cables.

Most video production setups that use ethernet for video connectivity require CAT 5e cabling or greater because regular Category 5 cabling only supports up to 100 Megabits per second of data transmission. CAT 5e supports a full gigabit, or 1,000 Megabits of data transmission.

Ethernet cable bandwidth.

Ethernet connections are easy and convenient to use for a variety of applications. For one thing, you can use network switches to connect many devices together over your network. You can also connect your internet router to a network switch to get all devices connected to your network. This gives all devices on your network internet access and the ability to send and receive video from anywhere on your local area network.

NDI is a popular IP video production standard you can use to accomplish this. Ethernet cabling can also be used to power devices such as cameras and lights with a power over ethernet (PoE) enabled network switch to power devices that support PoE power.

Audio cables

Audio cables are important for any

livestreaming project. The most common audio cables include XLR and TRS varieties.

THE BASICS OF LIVESTREAMING

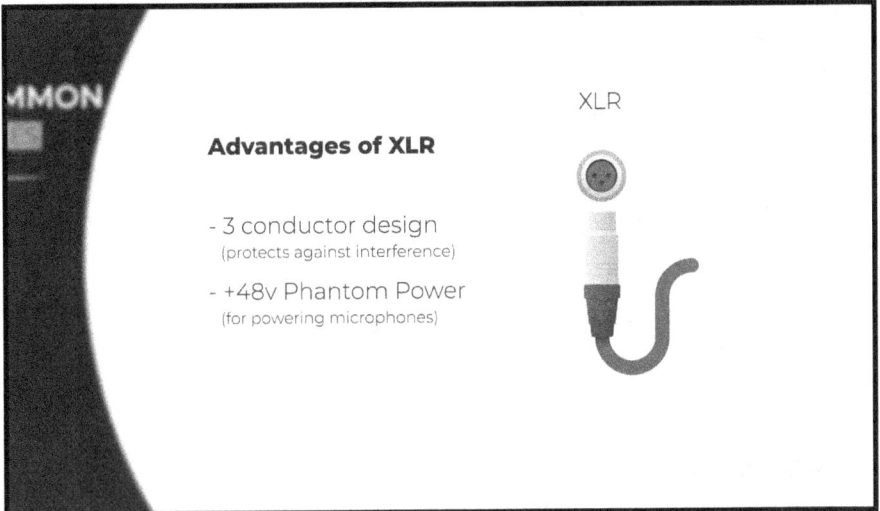

XLR cables.

In general, XLR cables have two main advantages over TRS. First, XLR cables have the required three-conductor design to support balanced audio where the signal protects against interference over long distances. The longer your cables are, the more important balanced audio is.

However, TRS cables can also provide balanced audio if the cable is connected between a balanced TRS output and balanced TRS input.

Second, XLR cables are also able to provide 48-Volt phantom power to microphones that use this feature. Your audio mixer must provide this power but the XLR cable is capable of transmitting it.

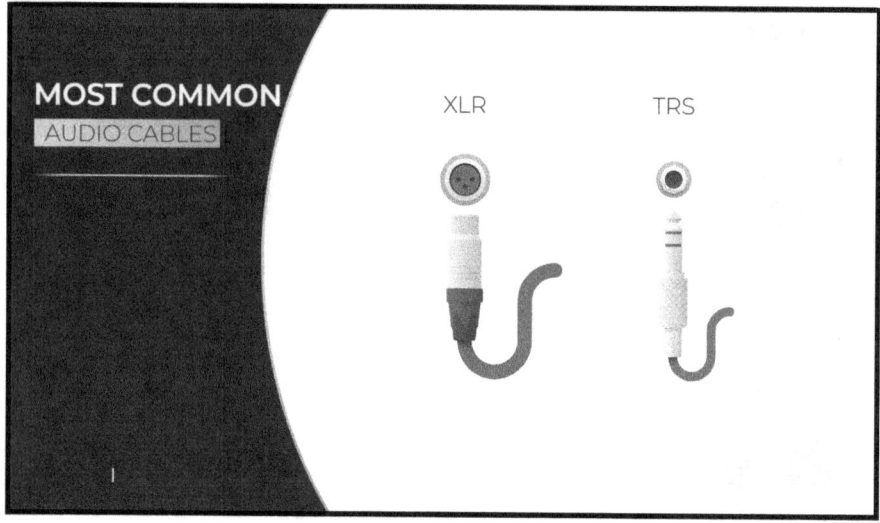

Typical audio cables.

Quarter-inch TRS audio cables are generally used for guitars and simple speaker set-ups, but XLR cables are preferred for many other applications such as professional microphones and digital pianos. Laptops and smartphones use 3.5 millimeter audio cables.

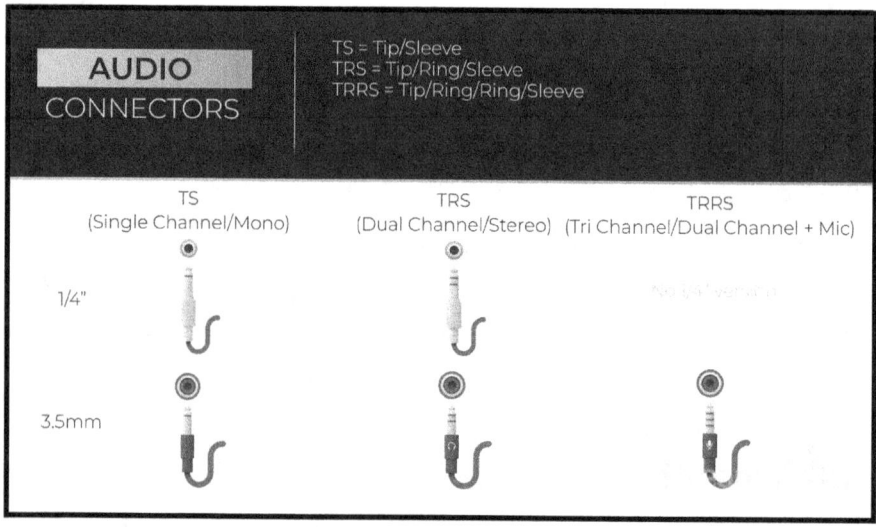

Audio connectors TS, TRS, and TRRS.

There are three types of audio jacks: TS, TRS, and TRRS. These stand for Tip/Ring, Tip/Ring/Sleeve, and Tip/Ring/Ring Sleeve. TS audio connections are mono, meaning they do not support left and right audio. TRS audio connections support stereo sound, while TRRS cables support stereo audio, along with a microphone connection. Both TS and TRS jacks are available in a 3.5 millimeter size, as well as quarter inch sizes. However, TRRS jacks are usually only available in 3.5 millimeter sizing.

Now you know more about the most important cables used for live video production.

13 HOW TO ADD GRAPHICS TO YOUR LIVE STREAM

Did you know that it's easy to add graphics into your live stream? Graphics can make any livestream more engaging for viewers. Simple graphics can display titles and livestream topics. Advanced graphics can integrate live viewer donations, comments, and even interactive poll questions.

What type of graphics can I add to my livestream?

The most popular and basic graphic many people use during livestreams is called a lower third. Lower thirds are graphics that cover the lower third of a widescreen 16:9 video space. Lower thirds are generally used to show information such as a show title, the current time, and even social media comments. You can create a lower third graphic by using a PNG file which has a transparent background. The lower third graphic can sit on a layer above the main video. You can either include the text information inside of this PNG file, or use software like OBS, Wirecast, or vMix to customize the information on top of your lower third.

Example of a lower third and spinning corner graphic.

Other types of graphics include countdown timers, logos, custom transitions, virtual sets, and interactive layers of web-based information. Graphics like lower thirds and logos can be created with software such as Adobe Photoshop or the free PIXLR software. Graphics such as countdown timers and video transitions can be created with software like Adobe After Effects. You can also save yourself some time by downloading free graphic templates on websites such as https://pixelpro.io/graphics. Or purchase premium graphics packages on websites such as Envato.

Example of picture-in-picture mode with a social media comment.

What software can I use to add graphics to my live stream?

Free streaming software like OBS provides the ability for any Mac or PC user to add graphics to their livestreams. OBS organizes projects into scenes which can create customized layouts of sources. Sources can include your camera sources, but also graphics of many kinds. OBS users will often add graphics to various scenes to add lower thirds, animations, and introductory videos to fit their scenes.

Popular livestreaming software choices.

Advanced video production software like vMix, MimoLive, and Wirecast organize channels of overlays that are used for graphics. For example, you may have a lower third on your overlay channel 1, and a rotating icon graphic on channel 2. These channels are always on top of the main video which can be switched throughout your video production.

How can I create custom graphics for my livestream?

As mentioned previously, you can use Photoshop or PIXLR to create graphics and software like Adobe Photoshop and Premiere Pro for video graphics. These software tools will enable you to customize all the colors for your graphics and upload them into your video production software. New cloud-based video production solutions like Restream.io and EasyLive are offering graphic overlays in the cloud. Video production software like OBS, Wirecast, and vMix generally always offer the most flexibility.

Pixlr log-in interface for online photo editing.

Where can I get graphics for my livestream?

The StreamGeeks have a free pack of graphics that you can download at StreamGeeks.us/graphics. The pack includes countdown timers, overlays, and lower thirds you can use to enhance your productions.

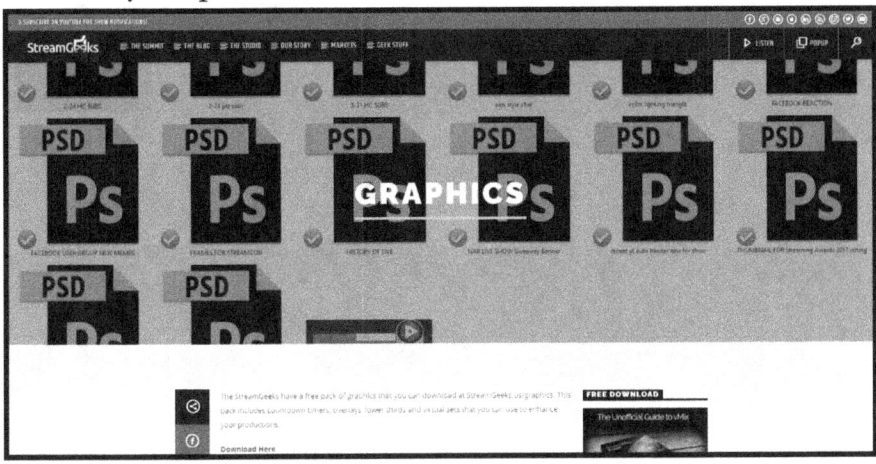

Free graphics from the StreamGeeks.

Let's take a look at a few of these files. The first is a selection of countdown timers. These are great for hosting a pre-show and letting your audience know when you will start your show. These MP4 files have either blue or green backgrounds which you can use with a chroma key filter to make them transparent. First, click the Dropbox link and download the files. Then add the files to your production software and add a chroma key. In OBS you can do this by right clicking the file and clicking the "add filter" function. Then choose "chroma key" and select the color you would like to make transparent. Now you can start the 10- minute countdown timer, exactly 10 minutes before your show starts.

Countdown timer.

Use the graphic overlay file to organize a video camera and a presentation space. This type of overlay can be used on a top layer above your camera and presentation files. For example, in OBS you might use this file twice in two unique scenes. One scene could show you next to a Powerpoint, and another scene might show you next to a video.

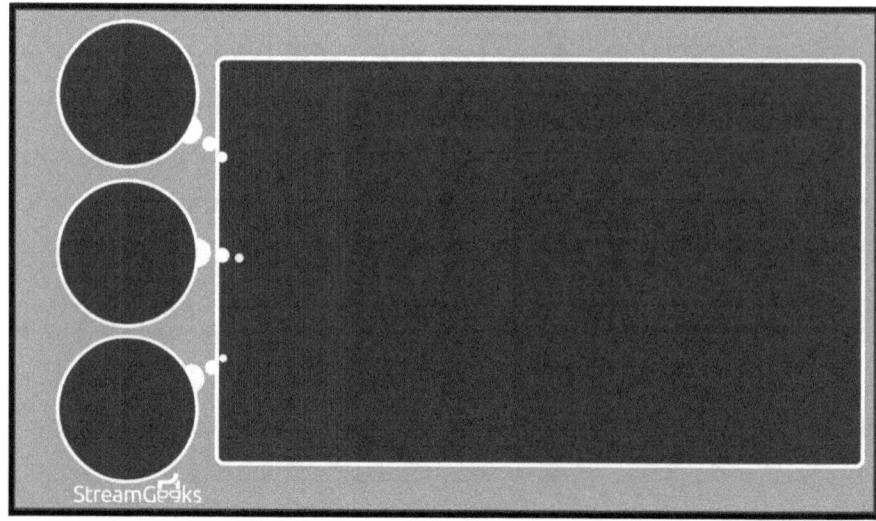

Example of a live show overlay.

Finally, our pro graphics pack includes some templates you can use for Zoom video conferencing and other talk show elements. There are overlays you can use to spice up your productions when bringing in guests. You may customize these graphics in Photoshop or through Pixlr. We also offer an entire pack of virtual sets that you can download for free. Check out our videos on using virtual sets with OBS, Wirecast, and vMix to learn more about using them.

Download the free virtual sets here:
https://ptzoptics.com/free-virtual-sets/

How can I add interactive graphics to my livestream?

Interactive graphics are the best type of graphics for live streams. Some interactive elements are platform specific. For example, on Facebook you can overlay poll graphics that viewers can interact with. You can set up your poll questions inside of Facebook and have viewers offer instant feedback during the livestream. Another cool example is Twitch Extensions, which are customizable modules that you can overlay on top of your Twitch browser. e For example, the interactive module allows viewers to see the player's inventory playing a specific videogame.

Other interactive graphics might include data sources from relevant information such as the weather, live donations, or recent data from your stream. For example, you may show popular chat comments, recent subscribers, or new donations. The possibilities are endless and a great place to start is your preferred content delivery network such as Facebook, Twitch, or YouTube. From there, you can look into interactive services available from StreamLabs and StreamElements.

16 WHAT IS NDI?

NDI stands for Network Device Interface and it is a high quality, low latency, IP video transmission standard popular for video production. This video connection type was initially adopted by the livestreaming industry in software such as: Wirecast, vMix, LiveStream Studio, OBS, xSplit, and the NewTek Tricasters. Today, NDI is used in a wide variety of video applications including broadcast, distance learning,and video communications. There are so many software and hardware integrations it's easy to use NDI with almost any project that uses video on a computer.

Most NDI users start their journey with NDI by seeking a new way to connect video sources beyond the traditional HDMI and SDI cable types. Getting started can be as easy as downloading the free NDI tools. Over the years, NDI tools have grown to include NDI viewing applications, screen capture software, virtual webcams inputs, and remote camera control options. There are even apps for iOS and Android that can turn your smartphone into a camera or presentation tool. NDI Tools allow anyone to get started using IP video and easily leapfrog old technologies that used to require expensive capture cards and hardware video switchers.

THE BASICS OF LIVESTREAMING

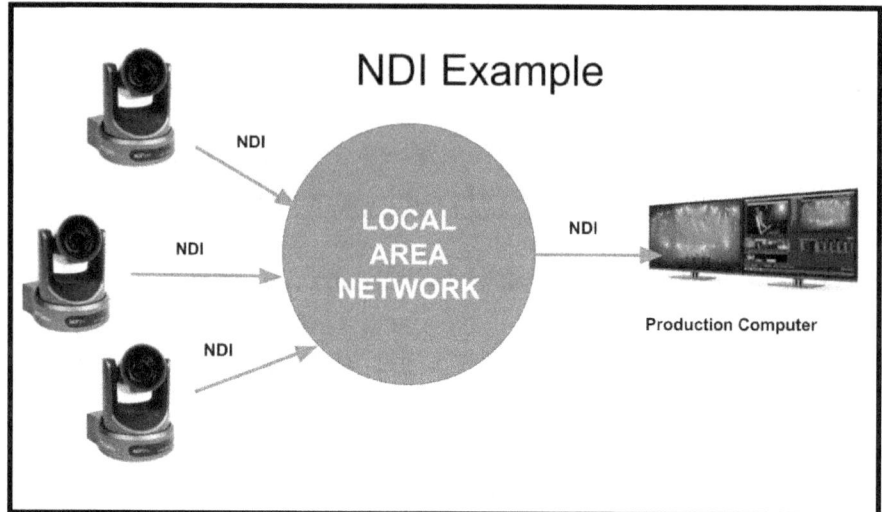

Example of NDI-enabled PTZOptics cameras connected to a local area network. These cameras are then discoverable with video production software like vMix.

While IP video is not a new concept, NDI has brought ease of use and flexibility to the complex world of network video. At its core, NDI has made the ability to quickly discover, send, and connect IP video sources incredibly easy and reliable. Over the years, NDI has also released new connectivity options such as NDI HX which stands for High Efficiency. NDI High Efficiency adds additional flexibility for bandwidth control when sending video over your local area network.

So how does NDI actually work? NDI requires a local area network, which is a network of computers connected together via ethernet cables. Networking equipment has become incredibly affordable and essential to modern communications, so you likely have a network in your home or office already. Networking equipment has revolutionized the way that we access the internet, communicate with the world, and connect our computers to other software and hardware. NDI is able to leverage standard networking equipment to allow you to send and receive video from various sources on your network.

Luckily, little networking knowledge is required to start using NDI. If you use OBS, it's simple, just download the NDI plugin and start sharing the output of your OBS production on your network with another computer. If you use vMix, try searching for NDI sources on your network that you might want to incorporate into your production, like a smartphone with the NDI app running. There are so many ways to use NDI with thousands of software and hardware configurations.

17 WHAT IS A PTZ CAMERA?

What is a PTZ Camera? It's become a buzz word in the tech industry, but many people may not know what the term PTZ camera means. PTZ cameras are pan, tilt, and zoom video cameras that allow an operator to control the camera remotely.

A tripod-mounted PTZ camera used to capture a rocket launch for NASA.

What does PTZ Mean? PTZ cameras can pan horizontally, tilt vertically, and zoom in on a subject to enhance the image quality without digital pixelation. Camera panning movement moves horizontally across a space, camera tilt movements move vertically up and down, and Zooming movements enhance the view from a camera with optical or digital zoom.

Why should I use a PTZ camera? PTZ cameras are ideal for video projects where you want to remotely operate a camera. If you are multitasking, using livestreaming software or other recording applications, camera operations can be automated using PTZ cameras. They're ideal when you want a single person to operate more than one camera or a regular camera just takes up too much space.

Are there different types of PTZ cameras? PTZ camera types are generally classified by their optical zoom and video output options. "Optical zoom" is a camera feature that allows you to enhance your image without pixelation as you zoom into an area. "Video outputs" are connections that you can use to bring the video and audio from the camera into your system. PTZOptics cameras, for example, come in 12X, 20X, and 30X optical zoom options and their video outputs include USB, SDI, and NDI connections.

PTZ cameras come in various lens types. Different lenses allow operators to zoom in close to the subject(s).

What are PTZ camera systems? PTZ cameras systems are complete video packages with the ability to record and/or livestream all necessary audio and video for any project. PTZ camera systems can be put together with Mac, PC, or dedicated hardware and they can include networking hardware to control cameras over an IP connection. Some even have the option to power cameras using Power Over Ethernet. PTZ cameras system may also include a PTZ joystick controller and other video production control interfaces.

19 WHAT IS A SDI CAMERA?

A SDI camera is a camera that has a SDI video connection. SDI video connections have a locking connector that can connect one end of a video device to another. SDI cables can connect devices such as a recorder, a production system, or a streaming encoder. SDI cameras can output video using a SDI cable connection.

What does SDI Mean?

SDI stands for Serial Digital Interface and it's a type of broadcast quality cable used to connect video sources such as cameras and TVs to other video devices such as video switchers, production systems, and encoders.

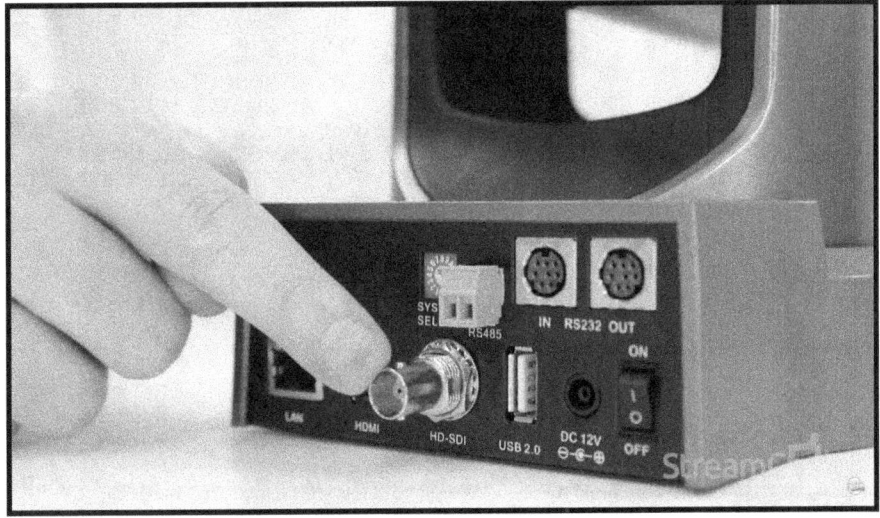

A PTZ camera's SDI connector.

What is a SDI camera used for?

SDI cameras are generally used to provide video inputs to production systems. Video producers will use SDI camera inputs to create a production that can be recorded or broadcast live. Video production switchers can be used to switch between multiple SDI camera inputs in order to tell a story and record live video content.

Is HDMI better than SDI?

HDMI stands for High-Definition Multimedia Interface. HDMI cabling is generally less expensive than SDI cabling but offers many of the same high resolution and frame rate capabilities. The main advantage of SDI cabling over HDMI is the long cabling distances that SDI supports. HDMI cabling is not typically reliable after 50 feet and requires acostly HDMI extension system. SDI cabling is generally used in more professional production systems, while HDMI cabling is mostlyused for consumer-based production systems.

Are all SDI cables the same?

SDI cables come in many different varieties. The first type is called Plenum vs. non-plenum cabling. Plenum-rated cabling is rated for use inside of walls and ceilings; therefore, many permanently installed cables that reach long distances require plenum-rated cabling to comply with fire safety regulations.

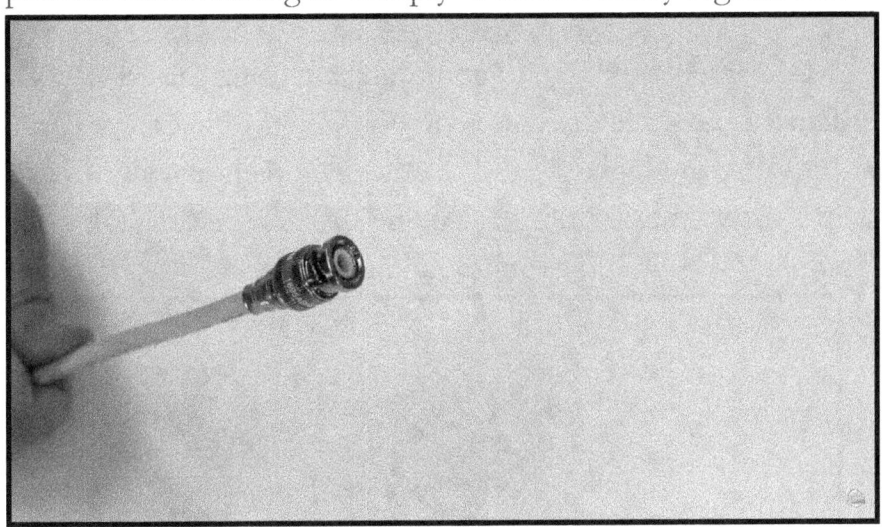

White SDI cable.

Common SDI cable quality types are SDI, HD-SDI, 3G SDI, and 6G-SDI. The original SDI cabling supports 480i video resolutions, HD-SDI supports 1080i/720p resolutions, 3G-SDI supports 1080p at 60fps, and 6G SDI supports 4K signals at 60Hz.

20 WHAT IS AN NDI CAMERA?

NDI cameras are able to communicate using the Network Device Interface or NDI protocol. They connect to a LAN (Local Area Network) and seamlessly integrate with hundreds of software applications including OBS, Wirecast, vMix, xSplit, NDI Studio Monitor, and much more.

NDI-enabled PTZ cameras.

What does NDI Mean?

NDI stands for Network Device Interface and it is a high-quality, video-over-IP standard developed by NewTek to enable video-compatible products to communicate, deliver, and receive high-definition video over a computer network ideal for live video production.

What is NDI used for?

Many video projects use NDI to send and receive video over IP. NDI has an auto-discovery feature which makes managing the video sources available on a network very easy. For example, a church may use NDI to send PowerPoint slides from one computer and receive them on another computer that is used for livestreaming. Another example would be setting up a display in an office to show a NDI source coming from a video production software like OBS. Any Windows or Mac computer can receive the NDI video stream and display it on a TV located in a facility.

What is a NDI camera used for?

NDI cameras often have PTZ (Pan, Tilt, and Zoom) features which take advantage of a NDI's two-way communication capabilities. In this way, NDI cameras can be controlled over the same ethernet cable used to send audio and video. For example, a PTZOptics NDI camera can use a single ethernet cable to power the camera, control the PTZ functionality of the camera, and send audio and video to a source on the network.

What is the difference between NDI and SDI?

SDI is a technology that has been around for decades. SDI stands for Serial Digital Interface, and the cable itself is capable of sending uncompressed video long distances. NDI is a much newer technology that uses the latest video compression methods to make sending and receiving high-quality video possible over standard computer networks. An SDI camera video feed can be converted into a NDI stream and sent over the network. A NDI video feed can also be converted into a SDI video output and plugged into a monitor.

What is NDI for OBS?

OBS, Open Broadcaster Software, is a free, open-source video production software available at OBSProejct.com. OBS is the world's most popular live streaming software and it is supported by a worldwide network of developers.

Palakis is a developer that has created a plugin for OBS which supports NDI. This plugin is available for both Mac and PC versions of OBS and it adds simple support for audio/video inputs and outputs over IP inside OBS. This allows OBS users to add NDI sources to OBS just like any source. It also allows OBS to transmit the main program video via NDI to other systems. It also provides a special OBS filter which can be used to output any OBS source via

NDI.

NDI sources in OBS.

How do I set up a NDI camera?

Most NDI cameras are plug and play when it comes to setup. NDI cameras can be plugged into any LAN (Local Area Network) and configured to operate with any software or hardware solution that supports NDI. Once a NDI camera is plugged into the network, it will show up as an available source on your network. Therefore, the NDI name that you give your camera will show up in any software or hardware solution when you click the "add NDI source" option.

20 WHAT IS A TALLY LIGHT?

A tally light is a light used in television and video production studios to signal to on-air talent that a camera is in use. Tally lights are located next to a camera that is connected to a video production system like OBS, Wirecast, vMix, or a NewTek TriCaster. A tally light will turn RED when the camera they are next to is live on air, and GREEN when the camera is in the Preview area of the video production system. When a camera is "live on air" this means that it is part of the "main program" being sent to viewers from the live streaming system. When a camera is in "Preview," this means that the video producer has the input in a preview window ready to be used for the next shot.

Wireless tally light from Tally-Lights LLC.

For video production systems that feature more than one camera, tally lights are used to signal which camera on-air talent should look toward. Without tally lights, on-air talent may not know which camera to look at. Some studios also provide on-air talent with a confidence monitor which is not always located near the cameras they are supposed to be speaking to. A confidence monitor displays the live video output to a stage or set where it can be viewed from. Confidence monitors allow people on-stage to quickly see what is being streamed during a live video production.

Tally lights are typically controlled by video production software systems such as OBS, vMix, or Wirecast but they can also be controlled by hardware systems like Blackmagic. Systems like one from Tally-Lights LLC. It can be plugged directly into your livestreaming computer via USB and integrated into the software you are using for video production. For example, the 8-camera tally light controller pictured is powered over the USB connection to your computer. It can then power up to 8 tally lights via 3.5 mm audio cables which send signals for the lights to turn on and off based on which input the livestreaming software is currently using.

Tally Lights Controller and tally lights connected with 3.5mm cables.

If more than one camera is running simultaneously, then both tally lights should remain on. In scenarios where multiple cameras are on at the same time in the same space, on-air talent should be trained to know which camera to look at. In some cases, a single tally light can be installed next to two cameras to reduce costs like the tally light pictured here installed between two PTZ cameras.

Tally lights can be wall-mounted, ceiling-mounted, or connected to video production gear through a variety of connection hardware. For example, the tally light pictured here connects to a tripod mount on top of a PTZOptics ZCam.

Some tally light systems are even used with monitors that feature built-in tally lights. These systems use a contact closure to know when the video production switcher should turn on a particular tally system.

Tally light shown on top of a PTZOptics Zcam camera.

Tally lights are a great way to make livestreaming equipment more usable for on-screen talent in scenarios where someone is on-stage or in a studio. For example, a pastor on-stage likes to know which camera is currently on so they know where to look during specific parts of a sermon. School broadcast clubs often like to have tally lights installed to let student presenters know exactly which camera to look at during morning announcements.

21 HOW TO BUILD A LIVESTREAMING STUDIO

It's time to talk about building your very own livestreaming studio. Your livestreaming studio should inspire your work and enhance your brand. You will need tos consider choosing a physical space, setting up lighting, sound, cameras, streaming software, and workflows. Once you have a good idea of how to create your own livestreaming studio, we will follow up with a tour of a couple of our live streaming studio builds.

First, how to choose a space? Choosing a space for your livestreaming studio starts with location. Choose a space that you and everyone in your team can easily get to. Iif you all work in a shared office, turn an under-utilized conference room into a fun livestreaming and content creation studio.If you work from home, take an extra unused bedroom or basement area and transform it into a live production studio.

Once you have chosen your space, start thinking about the backdrop for your set. If you have a blank canvas that's great, think about how you want to brand it. In the early stages of set design it's useful to use a Pinterest board to map out your ideas. Share the board with your teammates and get the creative juices flowing.

Think about painting a wall or choosing wallpaper. You can even have vinyl wallpaper printed and pasted on your wall to show anything your creative mind comes up with. Your backdrop is the first step to your livestreaming studio. You can also consider going with a green screen. Green screens offer many digital possibilities but they can also be limiting in the physical world. If you plan to use multiple cameras, you should consider building a real set.

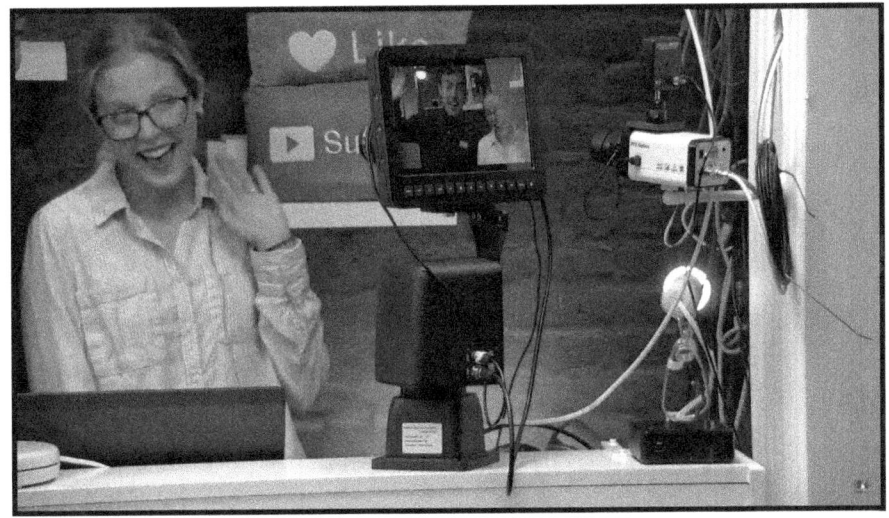

Example of a studio set-up.

Next, consider your presentation space. Do you want to have a table to show products in front of you, or are you going to speak directly to the camera for the entire show? In one of our studios, we have a D-Shaped Table with a TV monitor in-between. This is a stand-up table that allows us to keep the blood flowing as we present. Will you be sitting or standing in your studio? Maybe you want to get a standing desk that can do both?

In another one of our studio builds, we have a bar height table with stools. This allows us to stand or sit whenever we feel the need. It's really up to you how you design your set.

Next, consider lighting your set. Three-point lighting involves a key, fill, and backlight. Three lights are really a minimum for most professional studios. In our studio, we actually have five lights in order to provide light from multiple angles for each of our guests. In regard to lighting, consider choosing LED lighting which allows you to adjust the color temperature. The color temperature of your lighting can generally go from a warm color, around 3000 Kelvin, to a cooler color, around 7000K. Make sure to set all of your lighting to the same color temperature. Once you do this, you can set your camera's white balance to match.

One specific light that many people forget about is a backlight. A backlight is designed to shine on the back outline of your on-talent. It creates a halo effect around your head and allows you to stand out from your background. Lighting for your set makes a big difference.

Next, you should consider audio. If you like to move around a lot then consider choosing a wireless microphone. We like to move around so we use a Wireless Shure microphone system with headset microphones. You will have many choices for a microphone, but headset microphones are even better because they always stay right next to your mouth to capture crisp audio.

If you are always going to sit in just one space, you can set up a shot-gun microphone to pick up the area where you are sitting. Many people stream directly from their desk area and in this case, you can put a microphone on the table from where you are streaming.

When you are placing cameras in your studio you should think about framing. In fact, it's nice to have a camera on and set up as you are designing your studio. This way, you can place objects in-frame during your set design. Set up your livestreaming camera and zoom it into the space you want to be on camera. Then map out the objects that you will be place inside of your studio. Adding additional cameras will help you feature more detailed and engaging video content during your livestreams. Consider adding an over-the-shoulder shot to your production that brings viewers into your production. You can also set up a camera that is off to the side which can zoom into items in your studio. One popular camera is an overhead PTZ camera that zooms into objects that you have on your table. Another one of our personal favorites is a slider camera that can be used to show beautiful TV-quality sliding shots of objects in your studio.

Finally, let's talk about livestreaming software and hardware. It's amazing what you can do with a powerful Windows computer.

Demonstration of live streaming studio.

Notice that in our studio we have a custom Windows PC with a couple of PCIe capture cards inserted into it. This was

THE BASICS OF LIVESTREAMING

a wise choice because we have a powerful computer that we can use for a variety of tasks. Obviously, we install livestreaming software and you can see we have vMix here. But we also have Dropbox for sharing files, Google Chrome for accessing the internet, and even Adobe Premiere for creating videos in a video editing software program. We use a camera with 8 SDI camera inputs. You can choose SDI or HDMI video inputs for your camerasto connect directly into your computer. They can be used with most livestreaming software.

You may also want to consider getting some networking gear. A PoE networking switch can power many of the devices that you may want to use over a single ethernet cable.

22 WHAT IS SRT?

SRT is a video transport protocol designed to send high quality video over the public internet. SRT stands for Secure, Reliable, Transport. SRT can be used with many popular video production solutions including OBS, Wirecast, and vMix. SRT is used by video producers of any size to enable remote productions from around the world.

SRT is an open-source project started by Haivision.

Unlike NDI, which is designed for local area networks, SRT was designed for use over the public internet. This is achieved partly by managing a fixed amount of latency for each video stream. SRT video connections provide broadcast studios remote access to high definition video and audio that is usable for video production. For example, SRT is an ideal way to send video from reporters in the field who contribute remote video. c Broadcast studios receive that video mix it into a news production.

THE BASICS OF LIVESTREAMING

Remote production workflow.

SRT has made a name for itself by providing encryption that ensures secure transport of video with high production value. SRT can enable end-to-end AES encryption which is ideal for any content that requires protection. SRT protects against video jitter and packet loss even during bandwidth fluctuations from unreliable WiFi or cellular connections.

As SRT hardware and software has become more affordable, everyday productions at schools, churches, and state/local government agencies are using the solution. While SRT is used by the world's largest live production companies including Fox Sports, Comcast, and the NFL, the solution is open source and is also used by thousands of independent broadcasters.

SRT is helping to enable remote productions around the world. There are a few things you should know before you start using SRT.

First of all, SRT uses the public internet. Therefore, you can either set up a peer-to-peer connection or you can use a proxy server to connect. A peer-to-peer connection will require a little networking knowledge. For example, you will need to know your public IP address, make sure your router is set up with port forwarding, and configure your video production software to receive the stream. There are tutorials you can watch to walk through this process, like this one on vMix's YouTube channel. The second and easier way to set up SRT is to use a SRT MiniServer paired with proxy server capabilities. The application, which costs $30 per month, will receive SRT video feeds and convert them into NDI which you can easily use with OBS, vMix, Wirecast, and more.

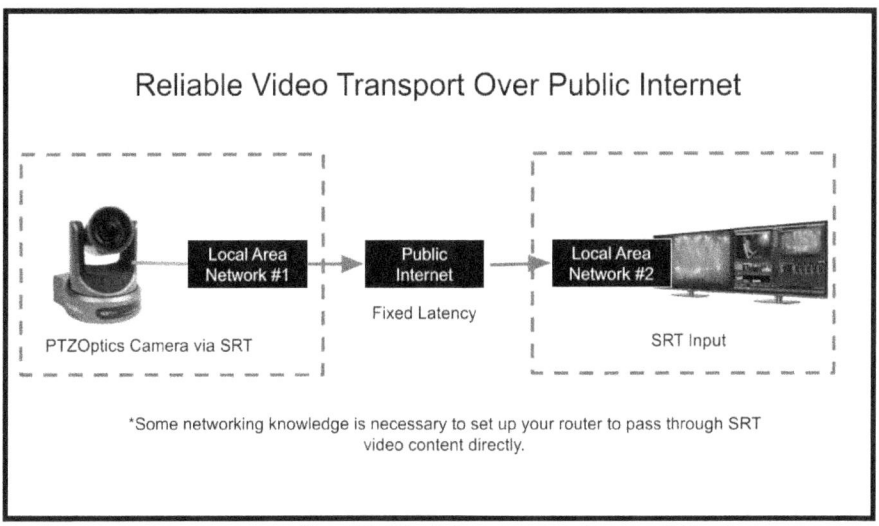

Common use case for SRT video over the public internet.

SRT is not ideal for two-way communications like Zoom or Skype. Rather, SRT is ideal for broadcasting in one-way communication scenarios like remote reporting and connecting video streams to a remote broadcast studio. One of our favorite use cases for SRT is remote production with a couple of 5G connected cell phones. The Urbanist produced a multi-camera tour of New York City using the Lorax Broadcaster App on iPhone 12 cameras. The cameras were able to send 2-3 Mbps quality video streams back to a production PC which tincorporated broadcast elements such as live chat, graphics, overlays, and more.

Another great example of SRT, is connecting two video production systems. For example, consider a trade show happening where a livestream is running. A company can easily set up a few cameras onsite, but it does not need to bring an entire video production system to the site. Instead, it can send each video and audio stream from the tradeshow floor back to a video production studio for broadcasting out to the world.

Note: When using the hardware encoding option in a software like vMix, you can usean NVIDIA graphics card to decode SRT video streams. When doing so, you may need to consider how many encoding channels are being used on your graphics card simultaneously. If your graphics card has a limited number of simultaneous encoding channels, you may need to use your CPU to decode SRT. Most Geforce NVIDIA graphics cards are limited to two hardware encodes per system. So if you are using hardware encoding for your recording and streaming, you cannot use it for SRT with a Geforce graphics card. If you have a P2000 or higher-end Quadro Card, you will have an unlimited number of encodes. Quadro graphics cards are only limited by the capabilities of the particular card.

23 CONCLUSION

Some video production professionals describe their attraction to livestreaming as a borderline obsession. Like a hobby, live streaming enthusiasts are drawn to the medium for the excitement it holds and the technology's ability to bring people together. Today, many people are introduced to the streaming industry by a friend or family member who livestream casually. Others are introduced to livestreaming in their professional careers. No matter how you discover the medium, the power of live streaming is clearly a global phenomenon.

In Asia, for example, livestreaming has become incredibly popular in almost every vertical, including shopping, live sports, fashion shows, video gaming, and more. A new term called "Shoppable Streaming" has emerged in recent years, to describe what small businesses in China are doing to sell products and services with live video. According to Latentview Analytics, 524 million Chinese viewers will experience shoppable livestreams in China this year. And, more than 100,000 brands have used it there, including Ralph Lauren, Clinique, Tommy Hilfiger, Lancome, Levi's, Louis Vuitton, L'Oreal, Burberry, and many others.

In the United States, livestreaming has grown significantly over the past 10 years. Perhaps the biggest growth increase happened after Facebook announced live streaming in 2016.

In the following years, Amazon has led the charge with its Twitch platform and its main ecommerce platform.

The StreamGeeks with a few friends at the 2018 NAB Show.

It is my hope that this book has helped you better understand some of the core technologies that enable livestreaming. This foundational knowledge should serve you well as you venture further into video production.

Checking out my other books to expand your knowledge on topics such as OBS, vMix, or Wirecast. All of my books are available on Amazon and on the StreamGeekswebsite via a free PDF download.

Feel free to email me directly at:
paul.richards@streamgeeks.us.

Cheers!

ABOUT THE AUTHOR

Paul Richards is a father, an author, and business executive leading his company in the exciting field of video communications. Richards is the author of multiple top-selling books that draw on his hands-on experience in the audio visual technology industry. As the Director of Business Development for HuddleCamHD and PTZOptics, Richards is the host of multiple online shows that feature his work on YouTube, Facebook, LinkedIn, and Twitch. Richards is the author of "The Virtual Ticket," "The Online Meeting Survival Guide," and "Helping Your Church Live Stream."

Paul is also the Chief Streaming Officer at StreamGeeks and teaches online courses for more than 50,000 students online on Udemy. His courses cover topics including live video production, online communications, and social media connectivity. On Mondays, you can find Paul hosting live streams on the StreamGeeks Channel. On Wednesdays, Paul appearson the PTZOptics channels talking about all things video production, cameras, and communications.

GLOSSARY OF TERMS

3.5mm Audio Cable: Male-to-male stereo cable, common in standard audio uses.

4K: A high definition resolution option (3840 x 2160 pixels or 4096 x 2160 pixels).

API [Application Program Interface]: A streaming API is a set of data a social media network uses to transmit on the web in real time. Going live directly from YouTube or Facebook login uses their API.

Bandwidth - The range of frequencies within a given band, in particular that used for transmitting a signal.

Broadcasting - The distribution of audio or video content to a dispersed audience via any electronic mass communications medium.

Broadcast Frame Rates - Used to describe how many frames per second are captured in broadcasting. Common frame rates in broadcast include: **29.97fps and 59.97 fps.**

Capture Card - A device with inputs and outputs that allow a camera to connect to a computer.

Chroma Key - A video effect that allows you to layer images and manipulate color hues [i.e., green screen]

Cloud-Based Streaming - Streaming and video production interaction that occurs within the cloud, therefore accessible beyond a single user's computer device.

Color Matching - The process of managing color and lighting settings on multiple cameras to match their appearance.

THE BASICS OF LIVESTREAMING

Community Strategy - The strategy of building one's brand and product recognition by building meaningful relationships with an audience, partner, and clientbase.

Content Delivery Network [CDN] - A network of servers that deliver web based content to an end user.

CPU [Central Processing Unit] Usage - The electronic circuitry within a computer that carries out the instructions of a computer program by performing the basic arithmetic, logical, control and input/output (I/O) operations specified by the instructions.

DAW - Digital Audio Workstation.

DB9 Cable - A common cable connection for camera joystick serial control.

DHCP [Dynamic Host Configuration Protocol] Router - A router with a network management protocol that dynamically sets IP addresses so the server can communicate with its sources.

Encoder - A device or software that converts a piece of code or information to then distribute it.

H.264 & H.265 - Common formats for video recording, compression, and delivery.

HDMI [High Definition Multimedia Interface] - A cable commonly used for transmitting audio/video.

HEVC [High Efficiency Video Coding] - H.264, one of the most common video formats:MJPEG-H Part 2

IP [Internet Protocol] Camera/Video - A camera or video source that can send and receive information via a network and the internet.

IP Control - The ability to control/connect a camera or device via a network or internet.

119

Latency - The time it takes between sending a signal and the recipient receiving it.

Livestreaming - The process of sending and receiving audio and/or video over the internet.

LAN [Local Area Network] - A network of computers linked together in one location.

Multicorder - A feature of streaming software that allows the user to record raw footage or a camera feed to a file separate from the stream output. [more]

NDI® [Network Device Interface] - Software standard developed by NewTek to enable video-compatible products to communicate, deliver, and receive broadcast quality video in a high quality, low latency manner that is frame-accurate and suitable for switching in a live production environment.

NDI Camera - A camera that allows you to send and receive video over your LAN.

NDI|HX – NDI High Efficiency, optimizes NDI® for limited bandwidth environments.

Network - A digital telecommunications network which allows nodes to share resources. In computer networks, computing devices exchange data with each other using connections between nodes.

NTSC - Video standard used in North America.

OTT Streaming [Over-The-Top] - When a media service bypasses typical media distributors (ie. Facebook, YouTube, Twitch) to distribute content.

PAL - Analog video format commonly used outside of North America.

PCIe Card - Allows high bandwidth communication between a device and the computer's motherboard.

PoE - Power over Ethernet.

PTZ - Pan, tilt, zoom.

RS-232 - Serial camera control transmission.

RTSP [Real Time Streaming Protocol] - Network control protocol for streaming from point to point.

Additional Online Courses:

Join over 50,000 other students who are learning how to leverage the power of livestreaming. Take the following courses taught by Paul Richards for free by downloading the course coupon codes available at streamgeeks.us/start.

- **Facebook LiveStreaming** - *Beginner*

This course will take you through the Facebook Live basics and it has already been updated twice! This also includes using Facebook Live Reactions.

- **YouTube LiveStreaming** - *Beginner*

This course will take you through the YouTube Live basics. It also includes essential branding and tips for marketing.

- **Introduction to OBS (Open Broadcaster Software)**

This course will take you through one of the world's most popular freelivestreaming software solutions. OBS is a great place to start livestreaming for free!

- **Introduction to xSplit Software** - *Beginner*

This course takes you through xSplit which has more features that OBS but costs roughly $5/month. Learn how to create amazing live productions and make videos much faster with xSplit!

- **Introduction to vMix** - *Intermediate*

vMix will have you livestreaming like the Pros in no time. This Windows-based software will amaze even the most advanced video producers!

- **Introduction to Wirecast** - *Intermediate*

Wirecast is the preferred software for many professional livestreamers. Available for Mac or PC, this is the ideal software for anyone looking to do professional streaming.

- **Introduction to NewTek NDI** - *Intermediate*

NewTek's innovative IP video standard NDI (Network Device Interface) will change the way you think about live video production. In this course, you will learn how to use this innovative new technology for livestreaming and video production system design.

- **Introduction to livestreaming course** - *Beginner*

This course includes everything you need to get started designing and building your show. This course includes a starter pack of course files including: Photoshop, After Effects, and free Virtual Sets.

- **Introduction to livestreaming** - *Intermediate*

This course focuses on more advanced techniques for optimizing your production workflow and using compression to get the most out of your processor. This course includes files for: Photoshop, After Effects, and free Virtual Sets.

- **Helping Your Church Live Stream** - *Intermediate*

This course focuses on livestreaming for churches and houses of worship. We tackle some of the big questions about livestreaming in a house of worship and dive into the specific challenges of this space.

- **How to LiveStream A Wedding** - *Beginner*

This is a great course for anyone looking to start livestreaming weddings. It was originally designed for wedding photographers to add a livestreaming service to their existing portfolio of offerings.